CADOGAN

Michael Pauls & D

the travellers' guide to
MARS

Don't leave earth without it

Travel	2
Practical A–Z	7
Mars in the Night Sky	19
Touring the Planet	22
A History of Mars	52
The Space Age Begins	87
Marsology	104

CADOGAN GUIDES
published by
Cadogan Books plc
27–29 Berwick Street, London W1V 3RF, UK
e-mail: guides@cadogan.demon.co.uk

distributed in North America by
The Globe Pequot Press
6 Business Park Road, PO Box 833,
Old Saybrook, Connecticut 06475–0833

series directed by Rachel Fielding
text © Michael Pauls & Dana Facaros 1997
ISBN 1-86011-014-2

first published in 1997

All rights reserved. No part of this publication may be reproduced, stored in a retrieval system, or transmitted, by any means, electronic or mechanical, including photocopying and recording, or by any information storage and retrieval system except as may be expressly permitted by the UK 1988 Copyright Design & Patents Act and the USA 1976 Copyright Act or in writing from the publisher.

project editor: Rachel Fielding
proofreading: Samantha Batra, Dominique Shead
specialist reader: Richard L. S. Taylor,.
            British Interplanetary Society
cover design: Horacio Monteverde
creative direction: Horacio Monteverde
book design: Kicca Tommasi
picture research: Dipika Parmar Jenkins

production: Book Production Services
publicity: Antonia McCahon

photographic and picture credits:
Ancient Art and Arch.: pp. 55, 59, 62, 66, 67.
Courtesy of Julian Baum: p. 101.
Bridgeman Art Library: p.64.
Mary Evans: pp. 6, 8, 9, 19, 21, 53, 60, 72, 74, 78, 80, 81, 82, 86.
Hulton Getty Images: pp. 61, 70, 71.
The Kobal Collection: p. 83.
Courtesy of Ron Miller: pp. 4, 84.
NASA: pp. 4, 10, 13, 14, 18, 22, 25, 29, 30, 31, 33, 36, 37, 41, 43, 45, 46, 48, 50, 54, 87, 88, 89, 90, 91, 93, 94, 96, 98, 99.

Science Photo Library: pp. 27, 35, 39, 47.
Scala: p. 69.
™ & © The Topps Co. Inc. p. 86.
University of Arizona Press: p.77
Maps © U.S. Geological Survey: inside covers
Universal Pictures: p. 95.

Every effort has been made to trace the copyright holders and we apologise for any unintentional omissions. We would be pleased to insert the appropriate acknowledgement in any subsequent edition of this publication.

A catalogue record for this book is available from the British Library
Printed and bound in Great Britain by BPC Wheatons
The author and publishers have made every effort to ensure the accuracy of the information in this book at the time of going to press. However, they cannot accept any responsibility for any loss, injury or inconvenience resulting from the use of information contained in this guide.

About the authors

Cadogan's dynamic duo, seasoned travellers and authors of 30 terrestrial travel guide books, Dana Facaros and Michael Pauls, have taken hundreds of virtual reality trips to the Red Planet in the last few weeks and are still hoping to pick up a cheap flight in the next launch window and make real footprints in the dust. Should they not return within the requisite fourteen months, their next project will be the *Cadogan Guide to Hell*.

Acknowledgements

The authors are beholden to all sorts of people for this book, starting with our Rachel and Tom Vaughan for simultaneously coming up with the silly idea of writing it in the first place. And special thanks to the Old Folks at Home for helping it along: Marge Pauls, in whose attic much of it was written, Chris Malumphy and Carolyn Steiner not forgetting Chris' ancient original Macintosh, which came out of the bottom of the closet to save the day when the leading brand software gremlins struck. More thanks to Michele Ch'i, who blew in from Hong Kong to tell the Chinese side of the story.

Many sincere thanks to the talented and delightful staff of the Cleveland Public Library, for tracking down nearly everything we needed despite being in the middle of a major building restoration; also for finding my glasses when I left them on a bookshelf on the seventh floor, and for artfully concealing their amusement at requests for obscure pop song lyrics from an odd character who claimed to be writing a guidebook to Mars. And final thanks to Alex Pavlot of Pittsburgh for his thoughts on Martian digestive systems. Alex is a good kid.

The publishers wish to thank Dana and Michael for being Martian enough to take this project on; Kicca for continuing to be creative even at the end of Martian-length days; Dipika for orbiting the earth in search of the right pictures; Richard Taylor for setting us straight in record time and for wonderful explanations; Peter Hingley, librarian at the Royal Astronomical Society, for pointing us in the right directions; Julian and Richard Baum for helping us navigate the Net and for kind comments.

for Marge,
the interplanetary hostess
with the mostest

Contents

Travel   2
- Getting to Mars   2
- Getting Around   4
- Passports, Visas and Entry Formalities   5
- Returning to Earth   6

Practical A–Z   7
- Climate and When to Go   7
- Clocks and Calendars   10
- Events   11
- Food and Drink   12
- Entertainment and Nightlife   13
- Phoning Home   16
- Religion   16
- Sports   17
- Tourist Information   17
- Where to Stay   18

Mars in the Night Sky   19

Touring the Planet   22
- An Introduction to Aerography   24
- The Martian Poles   26
- The Northern Plains   29
- Vastitas Borealis and Acidalia   29
- Chryse Planitia   30
- Cydonia and Arabia   31
- Utopia and Arcadia   33
- Elysium Planitia   34
- The Equatorial Regions: Tharsis   35
- Valles Marineris   39
- Syrtis Major and the Smudgy Bits   2
- The Southern Craterland   44
- A Day Trip to the Moons   47

A History of Mars   52
- Martian Genesis: A Troubled Childhood, Pimples, and a Damn Good Whacking   47
- Mars Gets Noticed   54
- A Gangster on Mount Olympus   58
- Rome's Red Grandpappy   61
- Meanwhile in China...   62
- The First Prussian 64
- Astrology's Middle-Aged Boy   65
- Florentine Intermezzo   69
- A Prisoner at Last   71
- Made in Italy—the Martian Canals   73
- New Earth, New Mars   75
- The War God Awakes   78
- A Not-so-Lonely Planet: Cultural Exchanges in the Early 20thCentury   80
- War of the Worlds Part II: All's Wells that ends Welles   83
- The Postwar Era: Mars in a Dark Funhouse Mirror   85

The Space Age Begins   87
- Early Red Successes and Setbacks   86
- Early Mars Tourism: Postcards from Modern Mariners, 1964–1971   87
- Phildickian Mars: Realities and Paranoias   90
- Mars Invaded by Vikings, Tastes Chicken Soup   91
- The Photo of the First Martian   94
- Bad Trips: 1980 to mid 90s
- But Is there life or at least SLIME on Mars?   97
- 1996–97 Here We Go Again   99
- Long-term Stays: Colonies on Mars   100

Marsology   104
- Further Reading   104
- Further Listening   106
- Further Watching   107

'To consider the Earth as the only populated world in infinite space
is as absurd as to assert that in an entire field of millet,
only one grain will grow.' Metrododrus

Travel

Getting to Mars

It will be a while before you can ring up a few days before departure and pick up a cheap charter flight to Mars. Under the best conditions (and the best conditions, or 'launch windows', occur roughly every 26 months) the round-trip journey is approximately 309 million miles. It will take between six and seven months just to get there, even when tripping along at speeds of 59,300 miles an hour. NASA currently plans to send a pair of spacecraft (a lander and an orbiter, similar to Pathfinder and Mars Surveyor) to Mars in 1998 (the lander is to stop near the south pole, if you fancy a skiing holiday) while one of the main tasks of the orbiter is to seek out likely places to hunt for underground water. So far the next probe, in 2001, is still being discussed, but if all goes well 2003 may be the year of the first return trip (a banner year to go: in August, Mars will be closer to Earth than it has been for the last 700 years). If not, it would be postponed until the next opposition, in 2005. As for manned missions, one isn't even planned yet. So far, the robots are doing just fine, and NASA has long believed that sending people would be too difficult and too

> the round-trip journey is approximately 309 million miles.

Portrait of Mr Golightly experimenting on Mess Quick & Speed's new patent high pressure Steam Riding Rocket drawn by C. E. Madley c. 1828

expensive—though with the current flurry of proposals for a major Mars colonization effort, this may change.

As for prices, they're going down all the time; luxury limousines like Viking 1 and 2 ($3 billion in 1997 dollars) are a thing of the past, and bargain basement fly-drive packages will be the trend of the future. Pathfinder and its rover Sojourner cost a mere $267.5 million, or 86 cents a mile—half the rate of a New York or London cab. Daniel Goldin, head administrator at NASA, hopes to get costs down to $77 million in the early 2000s (or a measly 24 cents a mile); at prices like that, you can't afford to stay home. The new parachute/air bag landing technique that worked such a treat with Pathfinder should make arrival fun–a bit like a bungee jump in a rubber ball.

Robert Zubrin, in his Mars Direct plan, notes that the fastest and safest way to go is when Mars is furthest away. By the strange calculus of space flight, a trip begun when Mars is opposite the Sun from us would take only six months each way—but it would require a 550-day wait on the planet in between, for a total mission of over three years. Zubrin estimates that once trips become commonplace, a passenger ticket will go for a mere $320,000. In the not-so-distant future, we will probably see faster, cheaper and more efficient ways to travel appear, which might be anything from nuclear electric rockets to sailing ships that catch the solar wind.

Getting Around

Pack every kilo you can squeeze in under the baggage allowance; on Mars, everyone is guaranteed to feel light on their feet. The relatively slight surface gravity makes every 100 terrestrial pounds into a mere 38 lbs, so carrying your grip will be much easier once you arrive. Available surface transport at the time of writing isn't very speedy; the only thing around with wheels right now is **Sojourner**, a little remote-controlled 'microwave oven on a skateboard', roving at a speed of a centimetre a second, or two feet a minute. It just might survive its first Martian winter, but you'll need to bring new batteries.

Soujourner: a 'microwave oven on a skateboard'

Meanwhile, the French space agency CNES is developing a faster mode of Martian transport: **hot air balloons**. Although their main job will be dowsing (each will trail a 'snake' full of instruments that will hover above ground during the day and drag along the surface during the night, searching for water) hitching a ride with one of these new-age montgolfières is the ideal way to tour the seven wonders of Mars. Some believe that some sort of airship will be the preferred mode of travel when Mars is settled. Though necessarily slow, they would be cheap, simple and require little fuel. They could even be filled with hydrogen, if Mars is kind enough to provide an easily accessible supply. Since the air contains so little oxygen, fires can't start, and we wouldn't have to worry about one blowing up like the Hindenburg.

Passports, Visas and Entry Formalities

Indeed, if a Man appears, no matter what rocket he is riding, there may be immigration difficulties, with even the threat of protest marches: MARS: NO ENTRY. EARTH-MEN, KEEP OUT

Gertrude L. Moore,
from Mrs Moore in Space

Astronomer Patrick Moore's mother seems to have had inside information, but these matters are unclear; check with the nearest Martian consulate. In spite of the current American monopoly on Earth to Mars transport, the US is a signatory of a UN treaty stating that Mars, like Antarctica, belongs to everyone. Of course this hasn't stopped some terrestrial entrepreneurs on the Web from selling off chunks of unreal estate (*see below: where to stay*).

If there's one recurring theme in all the literature about the prospects of exploring and settling Mars, it's that the enterprise will have more in common with the settlement of the New World on Earth than we might expect. Look for plenty of international and corporate wrangling in centuries to come, over matters we can't even imagine yet, along with a haphazardly emerging body of Martian law. Martian lawyers can't be far behind.

> the policy agreed to
> was the mutual right
> to shoot down any
> returning spacecraft

The night sky of Mars, with its two small moons and Earth gleaming like the evening star. Lucien Rudaux in L'Illustration c. 1930

Returning to Earth

Even if you spring for a $640,000 return ticket, this could be a bit stickier. To avert any chance of microbiological disaster on Earth, the Government of the United States agreed in writing with the Soviet Union in 1960 to a 50-year moratorium on any return trips from Mars. In fact, the policy agreed to was the mutual right to *shoot down* any returning spacecraft (this is probably not something you should mention to your agent when taking out travel insurance). Today, most scientists are convinced that the chances of any microbes or viruses even existing, let alone hitching a ride back with our spacecraft, are infinitesimally small. Most of the rest of us Earthlings, however, are equally convinced that one can't be too careful.

As it stands now, before NASA brings back any Martian rocks, (as, treaty or not, it envisions doing in 2005), it will have to state every possible worst-case scenario as part of an environmental impact statement for the US Environmental Protection Agency, which didn't even exist when astronauts retrieved the first moon rocks. There are scores of activist biochemists ready to keep NASA in court forever, on the chance that it might, despite all precautions, bring back a stowaway microbe or Ebola-type virus. If NASA is unable to reassure the sceptics, expect to have to pass through quarantine in an orbiting space station lazaretto (a possible new use for Mir?) where rocks and passengers will have

Practical A-Z

Climate and When to Go

to be thoroughly checked for viruses and other stowaway low-life forms.

There is also, of course, the matter of what effect our explorations will have on the Martian environment, which is worrying some scientists already. It may be too late: our microbes may have already arrived there with our Viking landers, despite careful sterilisation procedures. The terrible toxicity of Earthly germs to Martians was a theme not only of *War of the Worlds*, but also *The Martian Chronicles*. We can hope that terrestrial history won't repeat itself here, in the way that Europeans brought smallpox and the common cold to the New World and wiped out millions... but we may never know, or only know when it's too late.

Of the many little inconveniences of the Martian climate, most bothersome is perhaps the deplorable lack of **air**. The atmosphere is much thinner than the Earth's with a surface pressure barely 1% of our atmosphere, and this is **95% carbon dioxide**, dandy for a few bacteria but not so nice for us. Oxygen makes up 0.13%, and we suggest you bring along a good supply if you mean to do much breathing. There's also a small amount of nitrogen, 2.7%, as well as a few loose atoms of argon, neon, krypton and xenon, and equally small quantities of carbon monoxide, water vapour and ozone.

The planet's **magnetic field** is another problem—it hardly has one, so don't bother bringing along a compass. No one knows if this is a permanent or a temporary condition; studies in this area are still in their infancy, but it's suspected that changes in a magnetic field, or the lack of one, can have unpleasant effects on living things. Surface **atmospheric pressure** hovers in the 7 millibar range (Earth's is 1000), though this varies quite a bit with altitude. If you strolled on the Martian surface without your space suit you would have (briefly!) the curious sensation of freezing like a Popsicle while your blood boils.

Like Earth, Mars tilts on its axis, and therefore knows a passing parade of seasons—though the arrangement isn't quite the same. Mars' orbit is much more eccentric than ours, so the seasons are unequal and more inclined to meteorological extremes. Basically, the northern hemisphere has a more agreeable and temperate climate. Spring and summer up north are both longer than in the southern hemisphere, though down south summer temperatures can be as much as 50°F hotter. Summers last 178 sols (Martian days) in the north, while the south only gets 154 sols before the autumn chill sets in. And a pretty stiff chill it is; temperatures in the 183-sol southern winters can drop right off the thermometer.

Winters tend to be cold and cloudy, and because of the peculiarities of its orbit, Martian summers can be windy and plagued by dust storms, the worst coinciding with the southern hemisphere's short spring and summer, when clouds of powdery red dust hundreds of miles long are common. These can last for days and days, and then suddenly vanish. Some dust storms (as seen in 1971, 1973, and 1977) are global in scale, whipped by winds up to 73 miles per hour. Lately, since 1995, the Hubble Space Telescope photos have shown clear, unclouded skies, and no one knows yet if the dust storms or the clear skies are the more typical.

But no matter when you go, pack your long johns. If summer daytime **temperatures** can reach 60° F , the thermometer can drop the same night night to a bone chilling –90° F; in winter, expect no more than 0°F for a daytime high. –200° F has been recorded, and it might get down to –250° F at the poles. Remember, too, that the layered look is high fashion on Mars; Pathfinder learned that the temperature can vary up to 40°F within the course of a few minutes. On the positive side, you can leave your umbrella at home; mists and thin clouds may be common, but it hasn't rained for aeons. You won't have to worry much about the winds either. They can blow as hard as they like, but with the low atmospheric pressure there simply isn't much force behind them. We do predict, (and many science writers agree), that no matter how well-made and technologically perfect your space suit is, the dust will find its way into your shorts somehow.

We know what you're thinking—why not do the Greek Islands again instead this year? But enthusiasts of Martian settlement plans relish the very inhospitality of it all. They know that stories of the old days, when everybody had to walk around in space suits, will be something for the first settlers to brag about to their grandchildren. For they are convinced that we can change those conditions relatively quickly. The story of *terraforming*, the big dream that is making Mars so attractive as a potential destination these days, is on p.102.

For the record, in 1976, Seymour Hess, the meteorologist of the Viking Mission, issued the very first Martian weather report: 'Light winds at 15 m.p.h., shifting as any sensible wind is supposed to do. Temperatures Tuesdays ranging from a low of –122° F to an early afternoon high of –22 ° F and pressure of 7.70 millibars'. Martian weather is now charted periodically on a planet-wide scale by the Hubble Space Telescope. Quickly moving clouds of water ice zip over the poles. The enormous volcanoes, as far as anyone knows, are extinct, although two earthquakes were measured by one of the Viking landers. For more information, you can dial up the Mars Daily Weather Report from the Martian Sun Times at:

http://www.ucls.uchicago.edu//.

●

They know that stories of the old days, when everybody had to walk around in space suits, will be something for the first settlers to brag about to their grandchildren.

Clocks and Calendars

'Time moves more slowly on Mars....' the tourist brochures boast, and they aren't fooling. Up there, as on any foreign planet, you'll have to get used to entirely new ways of measuring time. Upon arrival, if you're worried about missing trains and appointments, the first thing to do is pick up an inexpensive **Martian watch**; these will work just like ours, only a little slower, to stay in step with the Martian day of 24 hours, 37 minutes. Note that while everyone does use the word 'day' informally, that period is officially called a **sol**, to distinguish it from the shorter Earth day.

Science has yet to deal with one of the biggest problems involved in the colonization of Mars—potential settlers will be facing a life in which they only get half as many birthday parties. What's more, a Martian could easily be a grandparent by the time he or she is 21. Quite simply, Mars takes 668.60 sols (or 687 Earth days) to travel around the sun. That's a year almost two years long, and winter can be a drag when it's two hundred below, and it lasts for six months.

As soon as permanent colonies are established on Mars, the parties involved are going to have to agree on a scheme for a **Martian calendar**. A number of these have already been proposed. Astronomer Patrick Moore thinks that 18 months of 37 sols each would do nicely. That makes 666 sols, or three short of the 669 needed. Moore suggests adding one day each to three of the months—but why not make them a three-day **intercalary festival** at the spring equinox instead? As Robert Graves once noted about ancient terrestrial cultures, 'intercalary festivals invariably degenerate into periods of unbridled license'. After a six-month winter, that is exactly what Martian colonists are going to need.

Moore did not attempt to invent names for his eighteen months. He preferred to leave that for the future, when he expects this question to 'produce the usual tedious international squabbles'. Perhaps the honour should belong to the first Martians themselves. There is one more loose end to tidy up. Remember that the year is really 668.60 sols, so a 669-day calendar will run slightly fast. Mars is going to need a **leap year**; if they subtract two days every fifth year, it should come out even—just as long as they don't subtract them from the intercalary festival.

Events

The native Martians may have some church or public holidays we don't yet know about; the data from Pathfinder suggests **rock festivals**, but no astronomers have ever reported seeing any fireworks. They probably wouldn't be much interested in commemorating the landing of the first Viking (20 July), or our discovery of their moons (11 Aug). Somewhat disappointingly, the only real events so far are **astronomical events**. Note that the Martian solstices and equinoxes are for the northern hemisphere (summer solstice, or the first day of summer up north is of course, the first day of winter in the southern hemisphere).

Earth date		Martian event (*terrestrial events in italics*)
1997,	13 Mar.	Summer solstice
	12 Sept.	Autumn equinox
1998,	6 Feb.	Winter solstice
	14 July	Spring equinox
	13 Nov	*Moon occults (passes in front of) Mars, seen from U. S.*
1999,	29 Jan	Summer solstice
	24 Apr	Mars at opposition
	31 July	Autumn equinox
	12 Dec	*Moon occults Mars, seen from Europe*
	25 Dec	Winter solstice (*Merry Christmas!*)
2000,	5 May	*Mars joins Mercury, Venus, Sun, Moon, Jupiter and Saturn in the 'great conjunction' that is bound to have people musing about the Age of Aquarius.*
	31 May	Spring equinox
	10 Aug	*Close conjunction of Mars and Mercury*
	16 Dec	Summer solstice
2001,	17 June	Autumn equinox
	11 Nov	Winter solstice
2002,	18 Apr	Spring equinox
	3 Nov	Summer solstice
2003,	5 May	Autumn equinox
	20 Sept	Summer solstice

Food and Drink

Science fiction often proves prophetic, and if this is true for matters of Martian cuisine the outlook is not good. Ray Bradbury, in *The Martian Chronicles*, makes possibly the first mention of a restaurateur on Mars. He is a ghastly, bad-tempered gringo, and the restaurant he opens is a hot dog stand (before this fellow sells many tube steaks, though, he goes mad and meets a bad end that he thoroughly deserves). If that's not disturbing enough, the Coca-Cola Company has already designed cans of its product that can be used in space. Please don't expect us to make any cheap jokes about Mars bars.

Mars is probably going to have a long wait before it gets its first Michelin-star diner. Right now, you're going to have a hell of a time even getting a decent cup of coffee—even in a partially terraformed atmosphere, water would boil at about 50°C. As tourism increases, perhaps standards will gradually rise, but until then all we can suggest is that you pack a picnic—really at least 450 of them, depending on when you start your trip and how long you mean to stay. The good news is that preserving food will be no problem. At Martian temperatures, a cheese sandwich will last several millennia, although you may want to make sure it's kept out of the ultraviolet rays.

Mars has water enough, but until we succeed in raising the temperature and melting it out of the permafrost, it will probably have to be transported from the north pole. Ice will probably still be easier to find than it is in European hotels. 'Always on the rocks, never on the house' is the slogan in the bars, but you'll probably only need one Phobos Fizz before you forget where you live—with the low atmospheric pressure, expect any alcohol to have quite a kick.

Entertainment and Nightlife

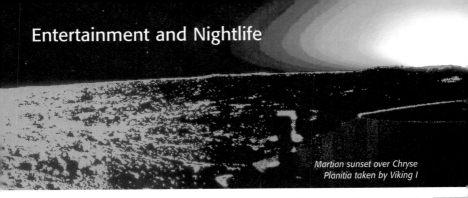

Martian sunset over Chryse Planitia taken by Viking I

Everyone says that Mars is dead after 9pm, but then, it's pretty much dead the rest of the day as well. Back in 1963, things were different. One of the terrestrial pop hits that year was the RanDell's classic *Martian Hop*, in which we find an eyewitness report that the Martians twist and stomp and that there isn't a single dance the Martians wouldn't do. Mars may be a very butch planet but there is some scant evidence of gay life: in *Stuart*, from the album *Beezelbubba*, a band called the Dead Milkmen hints at a secret correspondence between homosexuals on Mars and in Des Moines.

But face it, if you haven't brought along a special somebody to help you while away those long Martian nights (and they are 0.98% longer than ours), there won't be much to do but go out and look at the stars:

The Sky from Mars: First of all, if the idea of a salmon-pink sky disturbs you, you've probably bought a ticket for the wrong planet. Mars' thin atmosphere refracts sunlight a little differently, but scientists had still expected the sky to be a deep blue. Viking proved them wrong; indeed the sky colour was one of the first big Viking surprises, as the lander sent back pictures of a sunset that progressed from the normal pink to deep red. The colour comes from the presence of large amounts of dust particles in the atmosphere. On Earth, we see pink and red at sunset because of the greater thickness of atmosphere we look through when the sun is close to the horizon. Dust particles increase the effect here as on Mars; after a major volcanic eruption (like Krakatoa a century ago) we get prettier sunsets around the globe. Mars has the dust, and the pinkness, most of the time.

Note that there is still some difference of opinion about this. Due to technological limitations, space cameras can't really tell us much about true colours. When Viking 1 sent back its first colour shots, NASA processed them to make the sky blue, as they believed it was. Soon after, they decided that the dust they found in the atmosphere meant that it really had to be pink, and they adjusted the pictures accordingly, changing the tint of the sky (and everything else). Some people believe that the concentration of dust in the Martian air varies, and the sky colour along with it; therefore at times the pink may change to a purplish or yellowish tinge—or on the clearest days, even blue.

At night, thanks to the thinner air, the stars appear much brighter than they do on Earth (at least when there's no dust or clouds). There is another important difference. Mars' axis is tilted slightly more than our planet's—about 25° instead of 23°—and it points in a slightly different direction. Thus, the north pole points not to the star Polaris in Ursa Minor, as on Earth, but towards a spot near the bright star **Deneb in Cygnus**. As the night passes you will see the Swan (or Northern Cross) revolving slowly and prettily at the centre of the firmament. In the southern hemisphere, you will have a field of bright stars in the constellation Argo, including **Canopus**, to mark the pole.

Mars does have a brilliant blue-green morning star (or evening star, depending on its position): the Earth. Just as Venus, our closest inner planet,

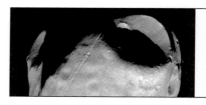

performs that duty for us, our own Earth announces the dawn or decorates the twilight for the Martians. It would go through phases, which would be plainly visible with field glasses, and anyone with average vision would notice it much brighter the closer it is to the Sun. You would occasionally be able to see our **Moon**, combining with the Earth to form a kind of 'double star'. **Venus** of course makes a second morning/evening star, and though smaller and fainter than we see it, the planet at least makes many beautiful conjunctions with Earth. To see Mercury at all, you would probably need a fairly good telescope or maybe field glasses, and a Martian almanac to tell you just when to look. **Jupiter**, on the other hand, puts on a fine show—along with the Earth it's the brightest object in the sky, excepting the Martian moons.

Of these, **Phobos**' low orbit keeps it so close to planet that it would not be visible from the polar regions. Even for viewers on the rest of the planet, it would usually be in eclipse, an absurd and lumpish coal-sack in the sky that blots out stars as it passes, appearing and reappearing only when rising and setting. **Deimos**, though smaller and correspondingly less bright, has a higher orbit and would be seen more of the time. Both these clownish moons do their best to provide some entertainment with their eccentric motions. Deimos, in its slow orbit, hangs in the sky for some sixty hours at a time, while Phobos, the sillier of the two, runs backwards, rising in the west and rushing through its phases as it scoots across the sky in only four hours. Both take great delight in **eclipsing the Sun**—Phobos does it over 1200 times a year, Deimos about 200. Since they are so small, though, they can only make annular eclipses, never entirely covering the Sun's face. Sometimes they eclipse each other, and sometimes they pass over the Sun together. In any case, neither moon is going to give you much light to see your way around at night, and neither is likely to be very conducive to romance.

> Phobos, the sillier of the two, runs backwards, rising in the west and rushing through its phases as it scoots across the sky in only four hours.

Between Mars and Jupiter lie the asteroid belts, and no doubt an observer on Mars would have an erratic passing parade of **asteroids** to contemplate. Some of them must have orbits that bring them close enough to Mars to be seen by the naked eye, though none of them would be very big or bright, only tiny dim lights hidden among the stars of the zodiac; you would have to know the stars well to recognize one.

Phoning Home

Although Mars Telecom has improved by leaps and bounds over the past couple decades, (each Mariner 4 photo, for instance, took 8½ hours to reach Earth and get processed) expect long pauses in your conversation. Messages from Mars, when the planet is at its closest, take about eleven minutes to arrive, even when travelling at the speed of light. When colonies are established, people will probably evolve some sort of Earth-Mars convention where both parties prepare their comments and questions in advance and transmit them at the beginning of the communication. When these are received, both will then make their replies to the other in a separate call, and they'll get them more or less simultaneously. It will be a lot like two people talking at the same time, but then that's what most of us do on the phone anyhow.

This, of course, assumes we'll have anything to say. Many sci-fi writers and colonization theorists seem to think that after people have been on Mars for a while they will lose touch with the Old World, or maybe lose interest. Then Mars and Earth will be like the Europeans and the Americans are now.

> ...expect long pauses in your conversation. Messages from Mars, when the planet is at its closest, take about eleven minutes to arrive...

Religion

Back in 1277, Pope John XXI decreed that, because God had the power to create whatever he liked, there could be more than one world in the heavens. The possible fossils of primitive organisms found in a Martian meteorite (one with a name like an address: Alan Hills 84001), prodded the Vatican to seriously consider its position on extraterrestrial life. Its main concern seems to be chronological. In short, if other life forms existed prior to the original sin of Adam and Eve, would they too need the redemption of Christ to be saved? There are (as yet) no plans to send missionaries on space missions, but it's not an idea the Church is ready to rule out.

There's cross-country skiing at the poles, and cliffs up to four miles high

Sports

Mars is an outdoorsman's dream (there isn't even any *indoors* yet). The possibilities for sport and recreation are endless, and so far pretty much unexploited— but think of all the fun you could have if you only weighed 76lbs instead of 200. There's cross-country skiing at the poles, and cliffs up to four miles high to really challenge rock climbers. Bring a deck of cards too. So far no sci-fi writer has dared to tell the awful truth about Martian exploration, but on the way there, it is most likely everybody is going to be sitting around for six months playing endless rubbers of bridge.

The slight gravity on Mars' surface does wonders for everyone's golf drives, but beware that the distances are correspondingly longer—the par 5 holes in the better courses are often $2/3$ of a mile long. If the truth be told, on Mars all lies are bad lies; most Martian golfers carry one wood, a putter and eleven different dust wedges.

Tourist Information

If the native Martians themselves are elusive, our own planet's cyberspace networks are jam-packed with information and chat. The Jet Propulsion Laboratory in Pasadena, in charge of the Pathfinder and Mars Surveyor missions, has a number of sites that are updated daily when data is rolling in (contact their **Centre for Mars Exploration** *http://cmex–www.arc.nasa.gov/* and related sites; the **International Mars Watch** (where professional and amateur astronomers share their news and sketches and photos of Mars) has a newsletter *(http://astrosun.tn.cornell.edu/marsnet/mnhome.html)*. **The Whole Mars Catalog** *(http://www.reston.com/astro/mars/catalog.html))* is a good site to start Mars-surfing, with links to all the NASA and other scientific sites, as well as all sorts of others. And you certainly won't have any trouble finding out about Cydonia and related subjects: there are nearly as many silly Mars websites as serious ones.

Where to Stay

Hopefully, a friendly and not-too-gruesome Martian will pop out of a crater and offer you bed and breakfast in one of the planet's legendary underground cities, because at the time of writing hotels are thin on the ground. You could probably squat, rather uncomfortably, under one of the Viking landers or the **Carl Sagan Memorial Station** (as the Pathfinder lander is now known) but the big question is, why rent when you can camp out on your own red patch? Now that Mars has been carved up by lines of latitude and longitude, three competing firms are selling off **Martian real estate** on the Internet.

All of these solemnly promise to present their land claims in the future to whatever governing body rules Mars (and if you think they're going to get very far, well, we can make you an even better deal on the oil fields of Uranus). 'Lands of the Universe' has already divvied up the planet into sections and townships, and offers four square miles for only $19.95, complete with a Certificate of Deed Registration and a 'personalized map'. Another company, 'The Martian Consulate, L.L.C.', offers 'prime Martian real estate...a perfect gift for Birthdays, Christmas, Hanukkah, Valentine's Day, Graduations, Mother's Day...'. People unfamiliar with the Internet might suppose we're making this up, but jests like these are probably nothing compared to what's going to happen when people actually start settling there. (*See* Long-term Stays p.102.)

> **the big question is, why rent when you can camp out on your own red patch?**

Mars in the night sky

You can read any number of books about Mars (there's a list on p. 104), and look at the photo-mosaics and atlases that have appeared since the Mariner and Viking missions gave us our first detailed view of the planet, but we would also strongly recommend going out at night to see the real thing, the only way we Earthlings knew it for thousands of years. Mars is a beautiful sight in the night sky (though totally uninteresting through field glasses or a small telescope!), and tracing its course through the seasons provides an object lesson in the complexities of the wandering planets' ways.

For over two thousand years, scholars would sit up late reading Plato's *Timaeus* after the doctors and abbots and professors had gone to bed. This great astronomical myth tickles parts of the fancy that Church doctrine, Aristotelian philosophy and even modern science could never reach. Timaeus sets out the neo-Platonic idea

Lucien Rudaux in L'Illustration

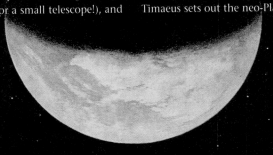

of the creation, and hints at the past Golden Age that ended when the firmament went awry—when the ecliptic, the path of the Sun, Moon and planets, separated from the celestial equator. We would say that this is simply a result of the tilting of the Earth's axis, but in the long history of neo-Platonic thought, it was the equivalent of the fall from paradise of Adam and Eve. Before it happened, the planets in their spheres progressed sweetly and evenly; while now they run mad races around the zodiac, now speeding up, now falling behind. Reminders of this old belief turn up everywhere in Western thought and literature:

Now Mars that valiant man is changed most:
For, he some times so far runs out of square,
That he his way doth seem quite to have lost,
And cleane without his usuall sphere to fare;
And euen these Star-gazers stonisht are
At sight thereof, and damne their lying bookes.

> Edmund Spenser,
> The Faerie Queene, Book VII; VII, 52

Since that mythical fall, as any astronomer from the Babylonians to Kepler would attest, Mars has been a problem. No planet's motions are so dramatic, or so difficult to explain. When Mars is close to the Sun, it seems to hang there for a long time, finally bursting out in the east and streaking across the sky, a second morning star for a season, rising a few minutes earlier before each successive dawn. From there it slows down gradually, growing brighter and redder, decorating the late night sky for months between its conjunction with the Sun and its opposition (the point when it stands exactly opposite the Sun, and rises in the east at sunset).

Then, surprisingly enough, it stops, and starts to move backwards relative to the stars behind it. This is the **retrograde motion** that perplexed the astronomers for so long, and forced Ptolemy to contrive his elaborate theory of 'epicycles' within the celestial spheres. All three of the visible outer planets—Mars, Jupiter and Saturn—show retrograde motion at their oppositions (when they are in a straight line with the Earth and Sun), but since Mars is closest to us, its 'loop' is the biggest and takes the longest to complete. Retrograde motion is not easy to explain, unless you can imagine yourself high above the plane of our solar system, looking down (see diagram).

Venus and Mercury, the inner planets that never seem to move too far from the Sun, shine most brightly when they are closest to it and passing behind it. This is because more of the planet's disc is visible; with binoculars you can easily see that Venus has phases like the Moon's, and that the phase is close to 'full' when the Sun is moving between Venus and us. For Mars and the other outer planets, the situation is quite different. They are brightest at opposition, when they are closest to us. Mars at its most brilliant can be fifty times as bright as the pale show it puts on when near the Sun; at opposition it upstages even Sirius, the brightest of the stars.

If all that wasn't complicated enough, Mars' slightly elliptical orbit throws yet another twist into the calculations. Earthly astronomers get their best chances to see Mars' surface when its oppositions coincide with its perihelion, its closest approach to the Sun. Regular oppositions come about every two years (the true interval can be from 764 to 810 days). **Perihelic oppositions** only happen every 18 years or so, and in our time they occur when the opposition is in or near Aquarius.

Mars' oppositional dance takes two and a half months to complete, and covers about half the width of the zodiacal constellation that hosts it. From there, the planet progresses slowly across the evening sky, appearing a little further west at sunset each night as the Sun rushes to catch it up. Then follows another long period of hibernation behind the Sun, and the whole two-year cycle begins again.

Imagine there is a bright star near Mars when the planet is at opposition (1). A month later, the Earth in its smaller orbit slips slightly ahead of Mars, and the planet now apppears to the right of that bright star (2). In another month (3) Mars is on the move again (from our viewpoint) and has left the star behind. Mars' retrograde dance depends on its position relative to the ecliptic. When it is slightly above it, it makes an upward loop (a); when below, the loop goes downwards (b). If opposition comes when Mars' path is crossing the ecliptic, we get no loop at all, merely an S-shaped shimmy (c & d).

Touring

the Planet

If this is your first trip to Mars, you'll probably want to follow the obvious tourist circuit: a ride to the top of Olympus Mons for the view, perhaps one of the package tours to Valles Marineris, and the obligatory stop on that crazy ol' moon Phobos. But there's a lot more to Mars than just these popular destinations. It's easy to forget just how big the planet's surface is. Though Mars may be much smaller than our planet, there isn't any ocean; consequently the total land area is about the same as Earth's.

Even at the relatively low magnification of the Mariner and Viking photos, we can see an amazing variety of landscapes: giant volcanoes, smooth plains, ridges and canyons, chaotic jumbled terrains, and every imaginable size and style of crater. Some Martian forms are so different from anything on Earth that we haven't even found names to describe them yet. All these are only the biggest and most prominent features. Even as we write this, the Global Surveyor is functioning well and sending back its first photos. And someday, when we get to walk around the planet and contemplate the little details, we'll see another Mars, on a human scale. Until then, no one can say how many odd corners and necks-of-the-woods will become famous for inspiring beauty, or exotic strangeness, or even—a certain charm.

We can't direct you to many of these now, though we certainly hope to be able to do so in future updates of this guide. And the sooner the better. ●

For some people, Mars is a planet only a geologist could love. And at this early stage of our explorations, a lot of the important work in deciphering its nature and its history is being done in this field. We know that much of the bedrock is volcanic basalt—the same that spews out of Mount Etna, that the Sicilians use to pave their streets. And we know that the rest includes plenty of iron. Mars' astronomical symbol ♂ is said to represent the iron shield and spear of the warrior god, and to astrologers and alchemists over the centuries that metal has always been associated with this planet . This is just as it should be, for it is the huge amounts of iron oxides in the rocks and dust that make Mars the Red Planet. This was confirmed by data from the Viking probes of the '70s, still our greatest source of information about nearly everything on the planet's surface.

Also like the war god of old, Mars loves extremes. Partly due to the lack of erosion, there are volcanoes over 80,000 feet above the imaginary sea level—almost 17 miles up, three times as high as Mt Everest. And the lows are lower; getting to the furthest depths of the Valles Marineris canyons would require a descent of over four miles below the surrounding plains. What Mars really lacks is mountains. Unlike the earth, the crust of Mars does not seem to be made up of moving tectonic plates, pushing up mountain ranges where they collide. But this does not mean that the surface is uniformly dreary and flat. Besides the great volcanoes of Tharsis, Mars has plenty of craters, with their circuits of 'walls' formed by the impact of meteorites, and tremendous lines of cliffs, some immeasurably higher than any on earth, and some longer than the Great Wall of China. You will also find modest hills and valleys in most parts of the planet, created out the plains by erosion of wind and water. Note that the most distinctive features as seen through the telescope won't necessarily be as striking from the surface, many of these are 'albedo features', and they stand out only because of the way their composition reflects light.

Mars doesn't have a sea level, so scientists have had to establish a more or less artificial substitute. If this

An Introduction to Areography

The Martian landscape is characterized by blocks that litter the landscape. The interest stimulated by these blocks depends on your point of view.

from *The Martian Landscape,*
NASA Viking Lander Imaging Team, 1978

standard, usually referred to as the **'datum point'**, really were sea level—if there were water under it—

(We Earthlings have already drawn up coordinates of longitude and latitude for the planet, with an arbitrary 0°

it would point out one of the many peculiarities of the planet's surface. Almost all of the 'seas' would be north of the equator, while the southern half would remain largely high and dry. The Earth is much the same, with its Pacific Ocean nearly covering a hemisphere, and the reasons for this condition on both planets are equally puzzling to scientists.

If water does reappear someday, it will leave two smallish 'continents' north of the equator, both created long ago by volcanic action: Elysium Planitia in the eastern hemisphere, and the Tharsis area in the west.

meridian in a part of the planet with no major features; thus we can talk about 'east' and 'west').

Getting to know Mars will require adding a few new words to your vocabulary. Areography for one, a newly-minted term defined as the study of Mars' surface (since the 'geo' in geography refers to the Earth, we couldn't logically use that term, could we?). There are already a score of other words to describe Martian landforms, which you can find in the glossary in the back of this book. As we get to know the planet better, more will undoubtedly need to be invented. ●

The Martian Poles

Along with Syrtis Major, the polar **ice caps** were the first features on Mars to be noticed by Earthly astronomers. Christiaan Huygens saw bright spots near the poles as early as 1672; others noticed them too, but no one seems to have made the obvious guess about their nature until Sir William Herschel, in the service of George III, did his extensive observations of Mars in 1779–83. Herschel observed the bright spots growing and declining over time, and determined that they were indeed 'frozen regions' like those at the poles of the Earth (it was while looking at Mars, incidentally, that Herschel discovered the planet Uranus, then lurking conveniently nearby).

Deciding whether to do your polar exploring in the north or south is largely a matter of what kind of ice you prefer. The permanent part of the northern ice cap is made up of mostly water ice, while down south you get frozen carbon dioxide, what we know as dry ice—the stuff that magicians and rock bands on Earth use to make instant steam for their special effects.

These ice caps work in much the same way as their counterparts on Earth, alternately growing and declining as the seasons change. At both poles, there is a small permanent cap, but it is the seasonal ones that make the grand show observable with telescopes from Earth. These are deceptive; though the seasonal caps cover the polar regions as far as 65° latitude in some places, the dry ice of which they are made is only a thin frosting. The opinions of the scientists differ; some believe the ice ranges from one to two feet thick on average, while others claim it's only a few millimetres in most parts—more of a long-lived frost. Mars seems to undergo a peculiar climactic shift every 51,000 years or so, when the polar caps change. The water ice cap gradually disappears at one and reappears at the other with a corresponding change in warm and cold winters; currently winters are milder to the north.

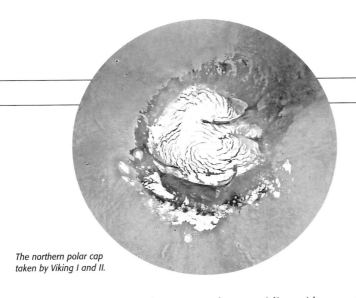

The northern polar cap taken by Viking I and II.

Polar exploration should not present any special problems. The terrain isn't particularly difficult, and the ice covering won't be enough to slow you down too much. As for the climate, it's a bit more freezing than everyplace else. The southern cap is the colder of the two; winter average temperatures of −280°F have been measured, compared to a balmy −200°F up north.

Mare Australe, the region of the southern ice cap, is markedly different from the moon-like cratered plains that surround it. Some small areas are flat and almost completely featureless, but for the most part the region shows long, swirling ridges and valleys formed out of what the areographers call 'laminated deposits'. Close up photos reveal distinct layers of these deposits, made of mixed dust and ice, and slowly scoured to their present shapes by wind and weather over the aeons. In some places these deposits may be over a mile thick, and the photos show them nearly filling up some craters on the edge of the polar plain. Interestingly, similar deposits have been noticed in other parts of Mars, at lower latitudes, leading to speculation that Mars' poles have shifted around a little through the ages.

The biggest of the southern polar valleys, **Chasma Australe**, starts near the pole and extends almost due east for over 200 miles, leading towards the most scenic part of the sub-polar region, including the magnificent **Promethei Rupes**, a crescent-shaped cliff broken by huge craters that curves for some 800 miles around the pole. As the southern winter commences, the ice cap does not spread out evenly. In the west, some parts as close as 84° S are free of ice year round, while to the east, between 0° and 270° longitude, the thin coating of ice expands as far as 65°, over 1000 miles from the pole. Here it covers most of Promethei Rupes, and some big craters behind the cliffs, such as **South**, **Main** and **Holmes**. Six Martian months later, the polar cap's retreat leaves behind patches of ice in the craters and valleys that last until late spring.

From space, the northern polar region **Mare Boreum** presents a surface that looks like wind-whipped curling waves on a sea. There are more valleys here than at the southern pole, and the prevailing winds have shaped them into these parallel curving patterns. Most are covered by water ice year round, with a thin dusting of carbon dioxide ice on top during the northern winters. There are no other major land features up here, only a few rare craters, of which the biggest is **Lomonosov**. Just to the north are some examples of the steep-sided crater type called a *tholos* on Mars. The real difference in the polar north is the dunes that surround the ice cap. It may not be quite correct to call them dunes, since they are formed by erosion rather than accumulation, and they probably aren't made of sand, but the resemblance to desert dunes in the Viking photos is striking. Martian weather patterns bring lots of dust to the north, along with condensing CO_2 that expands the ice cap at the start of northern winters. This makes the northern cap 'dirtier' than the south's, and over long periods of time it creates these thick deposits, which the wind erodes into dune-like shapes.

> the most scenic part of the sub-polar region, including the magnificent Promethei Rupes, a crescent-shaped cliff broken by huge craters that curves for some 800 miles around the pole.

The Northern Plains

Vastitas Borealis and Acidalia

Southwards from the 'dunes' that surround the polar ice cap (*see* below), this flat and featureless plain stretches clear around the planet. Parts of it are dark, others quite bright, though we aren't yet sure why. Most of the 'vastness' is also the lowest-lying ground on Mars. Once it may have been covered by a great sub-polar ocean.

There aren't a lot of craters up here, suggesting that this is one of the younger parts of the Martian surface. You will, however, see some peculiar ones: the so-called **pedestal craters**. Here the mass of debris thrown out from the centre by impact just sits and refuses to erode, resulting over the ages in a high platform above the surrounding plains. Some pedestal craters end up looking a bit like squat volcanoes.

Vastitas Borealis is only the northernmost part of an even vaster plain system that covers most of the northern hemisphere, in some places extending south of the equator. One part of it that stands out strongly in telescope views is **Acidalia Planitia**. On the ground this probably wouldn't look much different from the other northern plains, but its relatively dark colouring, combined with lighter-coloured ejecta from craters, gives it a curious mottled effect. Just to the west of Acidalia, on the way towards Olympus Mons, you could visit a region that boasts of some of the biggest **fossae** on Mars. These are formations of long, parallel cracks or depressions in the planet's crust; similar features can be observed on the Moon. Though relatively shallow, some of these cracks run for almost a thousand miles. On a map, you'll notice how the largest of them, **Tantalus Fossae**, **Mareotis Fossae** and **Tempe Fossae**, all point towards the giant Tharsis bulge and its volcanoes; the volcanic creation of Tharsis, or else the sheer weight of it, would have created the stresses within the crust that created the fossae.

Chryse Planitia

Chryse means 'golden', and the name for this otherwise nondescript plain is only another of Schiaparelli's telescopic fantasies. Right now, about the only thing to do on Chryse Planitia is look for the **Viking I lander**, which made Martian history when it landed here on 20 July, 1976. Today, this heroic little pioneer is as dead as the robot toys you bought for your kids last Christmas, but it surprised and delighted everyone by doing its job for six whole years, faithfully sending pictures and weather reports back to the JPL in Pasadena. Until the Americans come back to cart it off to the Smithsonian, you'll find it at 22.27° north latitude, 47.49° west longitude. Next to it you'll also see the first Martian celebrity rock, a six-foot boulder photographed by Viking and nicknamed **Big Joe**.

The landing spot was not chosen until Viking had already made a thorough photographic survey from orbit. It seemed like a safe enough spot, but safe also means dull; Earthlings wait-

first Martian celebrity rock, a six-foot boulder photographed by Viking and nicknamed Big Joe.

ing for the first pictures of the Martian surface were treated to a dusty, slightly undulating plain filled with stones and boulders. It reminded some observers of the Gobi Desert, others of the nastier parts of the Kalahari. This part of Chryse is also one of the lowest-lying parts of the surface, another consideration in its choice as landing spot; the greater atmospheric pressure helped slow down the spacecraft just a little bit more, slightly increasing the odds of a safe touchdown. If Mars is ever terraformed, this area might be one of the first small bodies of water to appear—Lake Viking, if you will.

On the Viking 1 mission, the plan was to stay in orbit for a while and photographically survey the surface; analysis of the photos would find the lander a good spot to touch down. Originally, NASA had wanted to land somewhere in Cydonia, but the surface there was so cracked and broken that mission control had to look elsewhere. At least, that's what they tell

All things considered, this still seems a bit low. But Cydonia is to us what the canals were to Percival Lowell's generation, and even after the first explorers have trudged all around it and over it, without finding any old coins or potsherds or ray guns, it is likely to hold a place in our legends and fantasies for a long time. The bottom line on the Cydonia hypothesis is

Cydonia and Arabia

the gullible public, according to true believers in Cydonia's Martian-made monuments (*see* p.95). This is the little corner of Mars that has caused all the commotion, and no doubt its wonders will be a prime tourist attraction even though the monuments turn out to be mere natural formations. A thousand years from now, Earthlings and Martians will still know the celebrated **face** (Cydonia believers see two 'faces', though the second isn't quite as distinct), along with the '**pyramids**' the '**city**' and the other formations we are imagining now.

Arthur C. Clarke remarked that his 'coefficient of scepticism' about the Cydonian monuments, once about 99%, has gradually gone up to 99.9%

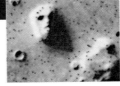

The Face

simple. Not only is the argument utterly laughable; it isn't even very interesting. The authors of this book are hardly sceptics: we hold many beliefs about past civilizations on Earth that would make an archaeologist go blue in the face. But if the explorers do find that Cydonia's marvels were constructed by intelligent beings, we would be happy to eat this book.

Cydonia Mensae, as it is properly called, covers the territory from 8° to 20° north latitude, and about 30–40° west. It lies on the boundary between Chryse Planitia and the broad, crater-filled region called **Arabia Terra**. It is also just above the datum point, and perhaps it was on or near the shore of the polar ocean back when its imaginary inhabitants were building their mighty monuments.

Arabia itself is a vast region, some 3,000 miles across, or roughly the size and shape of the United States. A plain heavily pockmarked with big craters, it looks more like the cratered highlands of the southern hemisphere than anything else up north. The biggest of these craters is **Cassini**, roughly where Illinois would be, and about the same size. The border regions of Arabia include some of the most interesting topography on Mars. Besides Cydonia, which (following the American analogy) would lie off the coast of the Pacific Northwest, there is the enormous valley **Ares Vallis**, further south, where California would be—a perfect spot for the landing of the Pathfinder mission, now officially known as **Carl Sagan Memorial Station**. Yes, the famous 'Rock Garden' is here, home to Yogi, Barnacle Bill, Scooby Doo and all the other rock stars that captivated the public's attention for a few minutes in the summer of '97.

Frivolous naming of Martian surface features, incidentally, is nothing new. When the Viking landers sent back their first pictures, the NASA team dubbed one long and more or less cylindrical rock the 'Midas Muffler'. It isn't surprising that the NASA wags try to put some life into the place. One of the Viking pictures, a boulder appeared with what seemed to be a clearly carved 'B' on it. The scientists started spending a lot of their free time looking for more inscriptions, and they found quite a few. As the NASA Imaging Team's report dryly noted, 'Commonly, the letters they discerned happened to be the first letter of their first or last names. Psychologists take note!'

Arabia's northern borders are mostly long stretches of really rugged terrain, the most dramatic of the Martian 'table lands'. These include **Deuteronilus Mensae**, **Protonilus Mensae**, and **Nilosyrtis Mensae**. The three combine to form a winding dark streak on the surface, and from the names, you can see how the early astronomers imagined a 'Nile River' flowing around the north of Arabia, and emptying into the 'great gulf', Syrtis Major. The imaginary river has two 'tributaries' in Arabia, **Auqukuh Vallis** and **Huo Hsing Vallis**, which get their names from Quechua (Inca) and Chinese words for Mars.

When Giovanni Virginio Schiaparelli was bestowing elegant classical names on the Martian locales he saw through his telescope in Milan, he must have fantasized these northern plains as the most delightful bower on the planet. He might have done better to call them Kansas and Nebraska, or maybe Nova Siberia, for you will be hard pressed to find such flat and

Utopia and Arcadia

uninspiring landscapes as these anywhere else on Mars.

Utopia's landmark, at least for people with our terrocentric prejudices, is the **Viking 2 lander**, which descended at 47.96° N latitude, 225.77° longitude on 3 September 1976, two months after the original Viking and 4500 miles to the east of it. The Viking photographed the first sunset on Mars, and beyond that, plenty of rocks. Unlike the pleasing variety of rocks shown at the Viking 1 site, these are all pretty much the same, and all profusely pitted (from gas bubbles in the cooling lava). They all might have come from the same source; scientists suspect they were a 'lobe of ejecta' from the **Mie** crater, 100 miles away

A low chain of north–south ridges called **Phlegra Montes** divides Utopia from Arcadia to the west. The Arcadian side does not appear quite so featureless, at least in close-up photos of its eastern half, rutted with fossae similar to those mentioned above; these too were probably created by nearby Tharsis.

Elysium Planitia

Like Utopia and Arcadia, Elysium is a name that would raise an understanding grin from any real-estate developer, but that is where the resemblance ends. Most of this planitia isn't even a plain, though the early astronomers could never have guessed it; the rise in elevation isn't even readily apparent from the Viking pictures. Elysium is the little sister of Tharsis, the smaller of the two great 'bulges' that distort the planet's surface, A rounded triangle in shape, it slopes gradually up to a height of three miles above the surrounding plains, making a platform for two of Mars' biggest volcanoes: **Elysium Mons** and **Hecates Tholus**. Elysium Mons, the larger and younger of the two, may be a midget compared to Olympus, but at some 40,000 ft above the datum level it would still easily out-top any mountain on Earth. Like Tharsis, Elysium has made its presence felt by creating networks of fossae in the surrounding area, but these are much fewer and smaller; the most noteworthy, **Elysium Fossae**, stretch east and west of the volcano.

You'll often hear this region referred to as the 'Tharsis bulge', and there's really no better word to describe it. Like Elysium, it is a huge, shallow bulge (or maybe a blister), one which rises up five miles and covers about 1/15 of the planet's surface. It seems to have appeared early in the planet's history, though we are a long

The Equatorial Regions
Tharsis

way from knowing how it came to be. Provocative guesses are not lacking. According to one, while young Mars was still cooling, it shrunk slightly, pushing up this part of the crust. Another notes that Tharsis lies roughly opposite the biggest crater on Mars, Hellas Planitia; the tremendous impact of the huge body—about 100 miles wide—that made Hellas simply pushed out a bit on the other side of the planet. This is intriguing, especially as the second-biggest crater, Argyre Planitia, just happens to lie opposite the other, smaller bulge of Elysium, but few geologists can be found who take the theory seriously. Most prefer the explanation that the bulges were formed by simple convection in the planet's mantle, the force that brings up molten material to make volcanoes and lava flows.(They seem to have been undergoing thermal uplift until quite recently). On Earth this happens all over the planet; in Mars' later history, for some unknown reason, it may have only occurred in two places.

The bulge is not so big that you would notice it from space; even on maps and photos, it isn't apparent unless contour lines are added in. Just the same, Tharsis affects almost everything that happens on Mars. As we have seen, it has shaped the landscape over vast areas of the planet. It helps shape Martian weather too, by influencing wind patterns. By distorting the planet's revolution, and the precession of its axis, it may partly determine long-range climate patterns.

Tharsis would be Mars' biggest northern 'continent' if there were oceans, stretching roughly 1200 by 1500 miles; its five-mile-high plateau makes a fitting platform for its four giant volcanoes, the biggest and youngest on the planet. The giant among these, **Olympus Mons**, is the tallest mountain yet discovered in the entire solar system, with a broad caldera at its summit 88,000 ft above the datum point—over 2½ times the height of Mt Everest. By volume, the biggest volcano on Earth, Mauna Loa in Hawaii, is only 1/100 the size.

On early maps of Mars, you may see this volcano appear as Nix Olympus, or the 'Snows of Olympus';

> Tharsis affects almost everything that happens on Mars. As we have already seen, it has shaped the landscape over vast areas of the planet. It helps shape Martian weather too, by influencing wind patterns.

this old name was given by Schiaparelli, who saw it only as a bright spot on the face of the planet, probably created by cloud cover. Just the same it was an excellent guess; if there are any gods on Mars they might well make this spectacular mountain their home, and in fact Olympus' slopes do often carry a bit of white—not snow, but carbon-dioxide frost.

cone itself—perhaps after a few centuries of terraforming this would be a good region for vineyards, like the slopes of Etna and Vesuvius. Is Olympus extinct? Most scientists think so, but we'll need a closer investigation to be absolutely sure. Some 19th-century astronomers saw bright flashes around Martian volcanoes, but as with the canals and so many other

Climbers should note that the easiest ascents seem to be from the east and southwest, where there are breaks in the cliff walls.

The diameter of Olympus' cone at the bottom is about 330 miles; compared to this, the biggest Earth volcanoes would be little more than pimples. Or to put it a different way, the base of Olympus' cone covers an area about as big as England, or Ohio. Vast rolling sheets of ancient lava flows extend to the east and north of the volcano, many times larger than the

matters, telescopic observation is not very reliable.

As imposing as Olympus may be from space, it's going to be difficult for photographers to get good surface shots to put on postcards. Most of the cone slopes up very gradually over its vast area, and from any distant point on the surface it would look not like a peak, but a vast, shallow dome.

Someone, someday will be the first to climb to the top of Olympus, to the rim of the caldera—because it's there, of course. But nobody is likely to be very impressed with the stories they tell. Getting to the top will be more of an immensely long and tedious hike than a climb, though the views will be terrific. With Mars' thin atmosphere, details of other volcanoes and craters in the distance stand out in much greater clarity than they do on Earth.

For a visitor on the surface, the most astounding feature of Olympus would be the escarpment around it, a buttress of cliffs that at its highest, on the southern edge, rises up over 18,000 feet. Climbers should note that the easiest ascents seem to be from the east and southwest, where there are breaks in the cliff walls. Some of the most difficult terrain will not be on the slopes themselves, but getting to them. Olympus is surrounded by an '**aureole**', a swath of finely ribbed, incredibly rugged land that in places is 400 miles wide; as a yet-unexplained volcanic artifact, this is another of the little mysteries currently keeping Mars geologists busy. Getting across the aureole would be murder, but fortunately there are wide breaks in it in the west and south.

Perhaps because of the presence of Olympus itself, this part of Mars may get some unusual weather; nearby Claritas Fossae (see below) seems to be one of the places where dust storms are prone to start. In any case, it does seem to have more weird erosion features than other parts of the planet. Among these the **yardangs** stand out: Yardangs are rippling, wave-like pat-

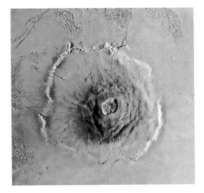

terns on the surface like those we often see on beach sand here on Earth—only here the ripples can be 30 miles long and a mile across; in some photos they grow so thick and close together that the planet seems to be sprouting hair. The regions south and southwest of Olympus, in southern **Amazonis Planitia**, have plenty of yardangs, along with dune-like formations and other oddities; in some places, you'll see hundreds of neat, straight parallel grooves etched into the surface by the wind; one of these covers an area of about a hundred square miles.

Not all the scratches in the ground are wind-formed. Near the western edge of Amazonis, in the cratery patch that separates this plain from Elysium, you might take time to visit the 300-mile long **Mangala Valles**, one of the largest 'channels'—valleys carved out by ancient floods—yet discovered.

Though Olympus may be the star attraction in these parts, it stands off by itself on the northwest fringes of the Tharsis bulge. Off to the east, you will find a number of *paterae*, smaller features originally believed to be craters, though these are probably also volcanic in origin: **Uranius Patera**, **Biblis Patera** and **Pavonis Patera** are the biggest. **Tharsis Tholus** and **Ceraunius Tholus**, small (at least compared to Olympus) conical mountains, are most likely volcanoes too.

At the top of the bulge, Tharsis' three other giants stand in a neat row 1,200 miles long; all three are nearly as tall as Olympus itself. **Ascraeus Mons** is the northernmost, and since it covers a much smaller area, it might be a somewhat more impressive sight from ground level. **Pavonis Mons**, in the centre, is just slightly shorter and as big around as Ascraeus; among the giants it is conspicuous for its dark slopes, and for the fact that its caldera lies exactly on the Martian equator. Flat-topped **Arsia Mons** in the south must be an even more remarkable sight from the surface; its caldera, 70 miles in diameter, covers nearly half the area of the entire mountain. Until Mariner 9, the three volcanoes were known only as North Spot, Middle Spot and South Spot, and some old veterans of the early Mars missions still like to use these terms.

In a terraformed Mars, if you let your imagination take a large leap, the vast calderas of the Martian volcanoes would gradually fill with water. Arthur C. Clarke (in *The Snows of Olympus*) imagines that in AD 2500 the Martian Winter Olympics can be held at any time of year atop—fittingly enough—Mons Olympus, and that a century later sailing in the caldera will be a popular sport in summer—though boatsmen will still have to watch out for icebergs.

Most of the early astronomers reported a broad, dark band around the equator of Mars. A number of the planet's most striking features are concentrated around its middle, leaving clues about Mars' origins and development that have yet to be fully deciphered. Foremost among them is this tremendous system of valleys and canyons, named after the Mariner spacecraft and the men who worked on it, those who first brought Mars' greatest natural wonder to light.

If Mons Olympus is the biggest mountain in the solar system, this is as far as we know the biggest and deepest valley. We would bet that when the first tourist hotel is constructed on Mars, this will be the place. Valles Marineris is four miles deep, 150 miles wide at its greatest extent, and an incredible 2500 miles long, one-fifth of the circumference of the planet. Any number of Arizona Grand Canyons could be quite easily lost inside it.

Some meticulous geologists have protested that 'canyon' is an incorrect term for the myriad cracks that form Valles Marineris. Canyons after all are

Valles Marineris

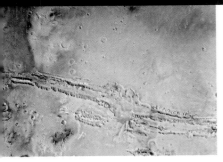

cut by rivers, and no place on Mars has any of these. The origins of the Valles remain one of the biggest Martian puzzles, but most scientists suspect that a fault line in the planet's crust is the major factor in their formation. It may be that the Valles are nothing more than a big 'crack' in Mars, caused by the pushing-up of the Tharsis bulge. Some who study Mars believe that the Valles are a result of tectonic plates in the crust moving apart, like some of the deep ocean trenches on Earth, but so far this opinion is still in the minority.

Valles Marineris is not a single canyon, but a network of them, some intersecting, others running parallel. Here are some of the most important components of the complex, running from east to west. First comes one of the strangest sights on Mars: **Noctis Labyrinthus**, the 'Labyrinth of the Night'. This chaotic jumble of connected canyons stretches for some 400 miles. Some of them run east-west, some north-south and a few at a slant; there are so many that they leave oddly angular blocks and triangles of raised territory in between; In the

> the ensemble resembles nothing so much as a message in Babylonian cuneiform writing, hurriedly etched on a clay tablet.

photos, the ensemble resembles nothing so much as a message in Babylonian cuneiform writing, hurriedly etched on a clay tablet. Seen close up, the Labyrinth presents some even more singular sights; tidy round depressions that look more like inverted cones than craters, mixed with ovoid and rectangular gouges and straight-line valleys. One section looks disturbingly like a human figure. It's a wonder that the Cydonia crazies haven't turned their attentions here yet; when they do we may have some more wonderful theories about Martian monuments of long ago.

Besides its strangeness, Noctis Labyrinthus has the distinction of occupying the very top of the Tharsis Bulge, six miles above the datum point—the highest spot on the planet, not counting the tops of the volcanoes. From here, the entire Valles system trends generally downhill. This fact, together with a close examination of the photos of this region, has given rise to some interesting speculation about the role of water in shaping the Valles Marineris long ago. In some places, water-formed channels do flow towards the canyons, and further west the parallel canyons generally converge. Even though geological processes did the work of creating these, some scientists believe it possible that when Mars still had surface water in ages past, it may well have passed through here, helping to shape the canyons into the forms we see today.

South of the Labyrinth lies a bright spot in the telescope's view: **Claritas Fossae**, as well as two small plains, **Syria Planum** and **Sinai Planum**. The Viking orbiters saw these plains covered in clouds.

Towards its eastern edge, the Labyrinth landscapes grow continually more chaotic, finally ending abruptly in a deep wide pit that marks the next stage of the Valles system. In this 400-mile stretch of straight, roughly parallel canyons, the most prominent are called **Tithonium Chasma** (on the north), and **Ius Chasma** (on the south). Close-up pictures of these show how tremendous landslides, some over fifty miles wide, have eaten away at the canyon walls, leaving wide swaths of rubble on the canyon floors. Parts of the walls themselves are almost two miles high. In Ben Bova's recent sci-fi potboiler *Mars*, the first astronauts to land on the planet risk their lives in an expedition into this rugged terrain, but are rewarded by finding a Martian life form—a sort of orange fungus. It may turn out to be a good guess on Bova's part. The deeper you go on Mars, the higher the atmospheric pressure, and it is at least conceivable that some hidden bottom in these canyons may go deep enough to allow water to exist in its natural state. Some of the canyons show layered deposits, as at the poles, suggesting that they were once under water. Bova imagined some volcanic source providing heat in the canyon bottom, creating a little microclimate that allowed life to survive after the lake dried up.

Tithonium and Ius converge at the broad but short **Melas Chasma**, which itself is connected to a series of equally impressive depressions to the north, forming a roundish bulge in the centre of the Valles system. The chasmas of **Hebes**, **Ophir** and **Candor** own fantastical names that perfectly match the fantastical panoramas. Ophir (named after the Biblical land that was the source of King Solomon's gold) gives the impression in the photos that it is collapsing, or eroding, into Candor; a smooth, 30-mile long, dune-like formation between the two provides one of the biggest and most unusual features of the area, which covers some 40,000 square miles—the biggest and baddest badlands ever seen by man.

Looking at close-ups of Ophir and Candor is like looking into a computer graphic of the Mandelbrot set; each successive magnification brings out new and uncanny erosion features, and each somehow seems related to the whole: wrinkled cliffs, delicately fluted valleys, canyons that branch into smaller canyons, and twisting ridges lined with rows of smaller, transverse ridges that slope down towards the valley floors. The lowest valley bottoms, four miles below surface level, lie hidden in shadow. If we ever settle Mars and start learning our way around, this is the place that the most adventurous souls will want to explore first.

Continuing westwards, Melas Chasma funnels into another giant canyon, the longest of the Valles system. **Coprates Chasma** carries on its neat and relatively straight path for over 500 miles; its walls, 60 miles apart in some places, show the same sort of landslide debris as Tithonium and Ius. Coprates got its name from Percival Lowell, who included it among his canals—the only one that was really 'there'. It splits in two near the end, leading into two ragged and twisting chasmas named **Capri** and **Eos**. Between them, they enclose a conspicuous dark spot, as seen through telescopes. This region doesn't have a name yet, but it's a perfect jumbled example of Martian 'chaotic terrain'. The darkness may be due to erosion having swept out parts of this area down to the dark volcanic bedrock.

Syrtis Major and

Why are some parts of Mars a darker shade of red than others? It may be a while before we get a conclusive answer to that one. A close inspection of the landscape doesn't offer many clues; the topography of the dark parts does not seem to differ consistently in any way from the light ones. The theories put forward so far fall roughly into two categories: either the mineral composition of the rock and soil of the dark parts is different, or else they are places where erosion is greater, exposing more of the (ostensibly darker) bedrock.

Whichever, the duskiest spot on Mars is unquestionably in the equatorial regions stretching eastwards from the Valles Marineris. Early astronomers often thought of these as seas, or 'gulfs' (*sinus*), and named them accordingly. Starting from the end of the Valles, the old Margaritifer Sinus seen by Viking cameras has resolved into a number of broad

the Smudgy Bits

expanses of rough terrain, of which the largest is **Margaritifer Chaos**. West of this comes a puzzling bright streak, and then an even darker region, **Sinus Meridiani**, that seems to be full of craters and otherwise quite similar to the brighter areas surrounding it. The smudge continues westwards through the **Sinus Sabaeus**, passing south of a part of Arabia dominated by a big crater that commemorates **Schiaparelli**, and the whole string of 'seas' finally debouches into the most conspicuous patch of ground on Mars, Syrtis Major.

Syrtis, named for the Gulf of Sirte in Libya, was noticed through telescopes even before the ice caps. Christiaan Huygens saw it 1659, and from watching it go round he easily deduced that Mars' rotation period is roughly the same as the Earth's. Later astronomers called the big smudge the 'Hourglass Sea' from its shape, though for us it's hard to see why. Syrtis is really just one of the older volcanic plains on Mars, and has no particular distinctions except for the fact that it is very, very dark.

Syrtis Major was also the first place where terrestrial politics intruded into Martian affairs. In 1869, a writer of popular scientific works named Proctor had made the first attempt at a systematic nomenclature for Martian features. Proctor named the major items after astronomers, and the old Hourglass Sea became the Kaiser Sea after a Dutch astronomer named Frederick Kaiser. About ten years later, Kaiser lost his chance at immortality when Frenchman Camille Flammarion made a new and improved map. He used some of Proctor's names but at that time a different Kaiser had just given France a very bad beating (in the Franco-Prussian War), and no patriotic Frenchman was going to leave that name on the map.

If you enjoyed your trip to the Moon, this may be the part of Mars where you'll feel most at home. The southern hemisphere offers craters galore, in landscapes that would be familiar to anyone who has ever and they show some of the same characteristics as craters on the Moon: surrounding walls, secondary craters and mounds of 'ejecta' outside, and occasionally central mountains. In areas where the older craters can be seen,

Southern Craterland

looked closely at our Moon; indeed many of the close-up photos from the Vikings are almost indistinguishable (in black and white at least) from comparable shots of our familiar satellite. There is one important difference—the rivers. Astronomers prefer to call them channel networks, but there is no doubt that once long ago they carried water down from the cratered uplands. From the patterns they show on the ground, it is believed that the flow of water was not gradual, but more like recurrent great floods. No one, unfortunately, is yet ready to explain how such a thing might have happened.

Mars has two 'families' of craters. The older ones, created in intense bombardments early in the planet's history, are mostly eroded away. New craters have been appearing ever since, including much of the southern highlands, we can guess we are looking at the oldest Martian landscapes, dating back 3.9 billion years or more. The smooth northern plains, where few craters at all are to be seen, must have been formed much later. Strangely enough a lot of them are the same size, about twelve miles across; the vast numbers of smaller craters the moon has are conspicuously lacking.

Seen from a distance, the landscape of the southern hemisphere is monotonous enough. Unlovely though it may be, however, if we earthlings are going to colonize Mars, we had better get used to it. If in a few centuries the Martian seas return, most of the dry land will be down here. The major features are two great impact basins, of which the largest is Hellas Planitia.

Something extremely big collided with Mars here long ago, disrupting the crust and allowing a tremendous lava flow to pour out and form the plain as we see it today. Hellas measures some 1100 miles across, making it by far the biggest impact basin on Mars. Its nearest competitor, the Mare Imbrium on the Moon, has a diameter of less than 700 miles.

If large amounts of water ever do come back to the surface, Hellas Planitia would be the only large body of water in the southern hemisphere—a series of small lakes or, if the water reached up to Martian 'sea level', a great roundish one, bordered on the east and north by the low hills called **Hellespontus Montes**. There are also a few small craters around the edges for landmarks, such as **Hadriaca Patera** to the east, near which are two long and probably very scenic valleys, minor versions of the Valles Marineris: **Harmaklis Vallis** and **Reull Vallis**. Right now, without the water, Hellas is one of the places where the big planet-wide dust storms have been seen to start. The region has proved difficult to photograph well, as it is often under a dust haze and mists. For all that, its depth (up to two miles below the datum point) and consequently higher air pressure makes it one of the Southern hemisphere's better sites for possible manned exploration missions.

Far to the east lies the other major feature of the south, **Argyre Planitia**. Like Hellas, Argyre is an impact basin, some 550 miles in diameter, and most of it is about a mile and a half below 'sea level'. It too has more than its share of dust storms—and also ice; the southern polar cap covers part of the basin in southern winters. Also like Hellas, Argyre is surrounded by rugged regions that pass for mountains: **Charitum Montes** on the south, and **Nereidum Montes** on the north. Both of these, however, look far more formidable through a telescope than they would from the surface of the planet. East of Argyre, the landscapes show a break in the craters, with hundreds of miles of territory covered with wrinkles and fossae that stretch from Tharsis halfway down to the south pole. The southernmost complex of fossae, and also one of the largest, is **Thaumasia Fossae**; while you're down there, you can take a look at one of the more interesting craters: **Lowell**, which has a huge inner ring.

Looking at the Martian craters from your spaceship, or even looking at the composite pictures (in the Atlas of Mars) made by the U. S. Geological Survey, you may be treated to one of astronomy's fond familiar optical illusions. Usually the craters look like proper convex hollows with high rims—but sometimes you blink and they suddenly stand out, like buttons, while the rims seem to have become deep moats. Up and down, Mr. Escher? Anyone who has ever spent much time looking at the Moon's craters through a small glass will recognize the phenomenon; just squint and persevere, and eventually the pesky craters will go back down where they belong.

The southern hemisphere

A Day Trip to the Moons

They're the joke of the solar system. Both tiny and shall we say, ill-featured, one of Mars' moons appears to go forwards, the other backwards—seen from the surface of Mars. Scientists classify the shape of both moons as an approximate triaxial ellipsoid, but we may be excused for thinking they look like a couple of potatoes—nasty old rotten potatoes, in fact, like those two that have been hiding in the back of your bin since the last astronomical epoch.

Though their presence was long suspected, the discovery of the moons had to wait for the big telescopes of the 19th century, and the intuition and persistence of an American astronomer named Asaph Hall. Hall worked at the U S Naval Observatory in Washington, and for the perihelic opposition of 1877 he talked his sceptical boss into giving him some precious telescope time to look at Mars. Most scientists of that time were sure that Mars had no moons—great astronomers such as Herschel had already looked for them in vain— but Hall went over all the gravitational calculations, and became convinced that if there were any, they would be much closer to the planet than anyone had thought. After many long nights, and brief teasing glimpses of unidentified objects near the planet's surface, Hall finally bagged his prey. On 18 August and Panic, the two attendants of Ares mentioned in Book XV of the *Iliad* (and nowhere else in classical literature; apparently the two were just a passing conceit of Homer's).

The confusion that followed would have been much more pleasing to the God of War than it was to Asaph Hall. First, his boss tried to grab part of the credit for the discovery, though he had

of that year, he was ready to announce to the world that Mars had not one but two moons (just as Jonathan Swift had written in *Gulliver's Travels* : *see* p 72). An English scholar wrote in soon after, suggesting the perfect names for the pair: **Phobos** and **Deimos**, Fear really done nothing at all. And after that, Hall's two assistants got hold of the telescope and soon announced that they had found yet two more moons, sending astronomers around the world on a wild goose chase that lasted for months.

For a century after Hall's discovery little else was learned about Phobos and Deimos. The pair were too small for even the biggest modern telescopes to get much of a look at, and they remained objects of mystery—especially the minuscule Deimos, which has attracted more than its share of attention from the world's crank theorists. Many UFO believers in the '50s held the opinion that Deimos was not a moon at all, but the grand mother ship, the base for all the flying saucers, slyly orbiting close to our neighbour planet so that we would not notice its presence. In the '80s an American supermarket tabloid published a story about a medium who claimed to have positive proof that Deimos was a base, and that Elvis Presley himself was alive and well and living up there as a guest of the aliens. (The King always did like his 'taters.)

Perhaps to shut up the cranks, Phobos and Deimos became prime targets for the Viking missions. Mission Control directed the Vikings to make near passes of the moons, though it took some tricky manoeuvering to get them close without crashing. Considering the time delay for sending instructions from Earth, they did an amazing job; the second Viking came within 17 miles of Deimos.

Despite the Viking photos, Fear and Panic remain as elusive as ever. Indeed, the photos have created almost as many new problems as they solved. Many of the rims of the moons' craters, for example, show obvious evidence of erosion, which for satellites much too small to have any atmosphere is disturbing. Some claim that micrometeorites or solar wind are responsible, while others maintain that Mars' atmosphere may well extend further out than we expect, far enough to create some very slight erosion on its low-flying moons. The photos also revealed some surprising surface features: long, straight grooves on Phobos, some which radiate from craters, and some which do not. Near the **Stickney crater**, the biggest feature on Phobos, these occur in neatly parallel rows. Scientists are still puzzling over them (Stickney, incidentally, was the maiden name of Asaph Hall's wife Angelina, who encouraged him to keep on trying to find those moons when he was ready to give up).

There are plenty of other puzzles, not least among them the moons' origins. Both are very dark and relatively non-reflective, unlike the bright and ruddy surface of Mars, and both are calculated to be much less dense than the parent planet. We don't even yet know how our own Moon was formed, and it may be a very long time before we can determine if Phobos and Deimos were created at the same time as Mars, or came along later. Two alternative explanations have been proposed so far. The moons may have been born from material torn off from Mars by the impact of a huge meteorite, or else they may have been strangers that just wandered in—lonesome chunks from the nearby asteroid belt that somehow came close enough to be sucked into Mars orbit.

The excellent science writer John Noble Wilford has compared Phobos and Deimos to the miniature asteroid that was home to *The Little Prince*. They are actually a bit larger than that. Phobos' diameters are 17 miles the long way, 13 the short way; in other words, you could ride a bicycle around it in two or three hours. For Deimos, the diameters are 10 and 6 miles, and some good stiff hiking would take you around it in an afternoon (there are as yet no bicycle rentals on Deimos).

You won't find much to see on either of them save **dust and craters**, but nevertheless a side trip to the moons could be the highlight of your trip to Mars. One attraction is some of the strangest and smallest horizons you will ever see; another, especially on Phobos, will be the spectacular views of Mars' equatorial regions—the most interesting parts of the planet. Phobos zooms around Mars in less than eight hours, and tourists could just sit and watch a constantly changing panorama of Martian landscapes flow by. Though it would be dark half the time, you would still get to see most of the planet in less than a sol. Deimos is further away, and since it moves with the planet you would have to wait a long time before other parts of Mars came in sight.

For all that, the moons' greatest lure might be good clean physical fun. *There isn't a lot of gravity*—about one-one thousandth of the gravity on Earth. You could jump into the centre of the mile-deep, 6 mile-wide Stickney crater on Phobos, land without hurting yourself, and then jump right back out again in a few bounds. The escape velocity from Phobos is about 13 mph, from Deimos even less; you could easily throw a stone off either of them. On Deimos, you might be able to jump right off, in which case you would become a third moon, sharing Deimos' orbit. Because of the lack of gravity, people have speculated that the moons may someday provide a convenient base for exploration of Mars and the outer reaches of the solar system; landing and taking off would be wonderfully easy. We are more inclined to think that if the Americans ever do land on Phobos or Deimos the first thing they will do will be to set up a pair of hoops a mile or so apart, and get up a game of basketball.

There is one problem with all these fantasies (besides the lack of air, of course), and that is the *regolith* that haunts both these moons. A regolith is nothing more than a coating of dust and debris, probably caused by the impact of meteorites, and Phobos and Deimos both have more than they can use. It is estimated that the regolith is at least 15 ft deep in most parts of the moons, and probably much greater, so it might be difficult getting any traction for jumping or bicycling or shooting baskets.

Don't let any of the Martian real estate web sites sell you any lots on Phobos, for this moon has no future. Scientists estimate that tidal forces will cause the moon to break up and/or fall into the planet sometime within the next 100 million years. Deimos, on the other hand, seems to be gradually expanding its orbit further away; someday it might get lost, or perhaps go back to join the rest of the space debris in the asteroid belts. ●

> Don't let any of the Martian real estate web sites sell you any lots on Phobos, for this moon has no future.

A History of Mars

Martian Genesis: A Troubled Childhood, Pimples,

Mars, along with the other planets in our solar system, was born about five billion years ago in a whirling frisbee of hot crud spinning around the sun. As the frisbee puffed out into a halo, much of this mass went hurtling off into space; the remaining lumps of junk, over a few million years, eventually became attracted to one another through gravity and loneliness, and embraced to form planets—in fact, they embraced with such passion that the rocks in the planets' cores melted.

As Mars cooled, some say, the molten rock became lopsided, forming two swollen lumps under the crust, the Elysium bulge and the Tharsis bulge. The molten lava and heat bottled up inside these bulges was vented through volcanoes; the worst of these, Mons Olympus, soon grew to be larger than England. The out-thrusting of the Tharsis bulge also left enormous fissures in the crust of Mars, including another Martian superlative: Valles Marineris, the biggest crack science has yet discovered in *anything*.

Mars was still only a tot—only a billion or two years old—when it was blindsided by a violent rain of meteors, scarring and pitting the entire surface with millions of craters. Some of these are enormous—the Hellas basin, 1100 miles in diameter, is the largest known crater on any planet. The

> *'...a kind of mythic arena onto which we have projected our earthly hopes and fears.'*
>
> *Carl Sagan*

and a Damn Good Whacking

impacts were so powerful that the *ejecta* (bits of Mars) went flying in all directions, becoming asteroids or mere space debris—some of which has turned up on Earth. Mars' tiny moons, the Laurel and Hardy of satellites, may even have started out as *ejecta* from the mother planet. According to some current thinking, the intense beating Mars took released gusts of heat that transformed elements into gases, forming an oxygen-rich atmosphere. There was also water: flood channels have left their unmistakable imprint across the surface of the planet, and perhaps glaciers, rivers and seas were present too.

This warm and relatively cosy Mars of *c.* four billion BC was sadly doomed from the start. First, because Mars is small (its mass is only 0.1074 that of Earth), its core cooled down quickly, depriving the planet of most of its precious magnetic field (our terrestrial magnetism is created by the difference in rotatation between the hot core and the rest of the planet). Without its magnetic field, Mars' surface was deprived of a shield from deadly radiation; its atmosphere fell prey to erosion from solar wind. Even more atmosphere may have been lost from the impact of asteroids like those that made the Hellas and Argyre craters.

Unable to retain heat in the thinning atmosphere, the water and what remained of the atmosphere froze and sunk to the surface, where the oxygen bonded to the rock of the crust and rusted it red, leaving only carbon dioxide in the Martian atmosphere. The frozen water was then warmed by the surface heat, only to seep deep down to form underground seas. The pressure, or perhaps the eruption of a volcano or shock of a colliding meteorite or comet would release the stored up water and cause cataclysmic floods; according to some estimates, if all of Mars' subterranean aquifers were to burst forth, they would coat the planet with a 1000ft-deep sea.

This cycle of flooding, freezing and seeping down occurred several times, and each time the water would freeze and sublime into the atmosphere, turning into snow that fell on the poles. The polar caps grew thick and heavy while much of the rest of the water was thrust deep underground. There it has remained, leaving Mars the dry red biscuit we know, moulded by winds and dunes of dust.

Or as Kim Stanley Robinson described it in *Red Mars*: 'the planet, taken in itself, is a dead frozen nightmare (therefore exotic and sublime).' For the last few thousand years at least, Martian history has been made only on another planet—ours—and Mars has had a much greater effect on us than we have had on it. Yet while we are forced to resume the story from a geocentric point of view, the careful reader may glean hints of what the Martians have been up to. And it's usually no good. They are nothing if not masters at fooling with our minds.

Mars Gets Noticed

There are times when Mars scarcely stands out among the myriad constellations in our night sky, and there are other times when it burns fifty times as bright, glowing like a red ember. Most often, it promenades respectably across the sky with the other planets, but now and then it appears to hesitate, and even reverse. One can only guess at the wonder, curiosity and fear Mars must have caused its first observers—but those observers lived untold aeons ago. By the 3rd millennium BC, the Egyptians already recognized that the odd behaviour of

Har decher ('the red one') at least followed predictable patterns.

To them, Mars was 'the Star of the East of Heaven', also described as *sekded-ef em khetkhet*, the 'star which journeys backwards in travelling'. But just where it fits into their religion and mythology is, as usual, something neither the Martians nor the Egyptians want us to know. One clue may be the fact that under the XIXth and XXth dynasties, Mars was called *Heru-Khuti* (or, as the Greeks wrote it, *Harmachis*). *Heru-Khuti*, sometimes pictured with a hawk-head and a triple crown that seems to be made of bowling pins, is familiar enough. He is usually considered one of the chief forms of the Sun God Ra; as such he had an important shrine at Heliopolis, and another one at Giza—the biggest monument to Mars or any other planet, nothing less than the famous Great Sphinx. It is impossible to say just how much Mars really had to do with it. The enigmatic Sphinx is partly a political monument, wearing the face of the Pharaoh who built it. Perhaps the astronomical connection includes a little of both the Sun and Mars: for the King as ruler, and as warrior.

The Mesopotamian cultures, as far as we know, were the first to see something sinister in the Red Planet. The Babylonians called it **Nergal**, the Death Star, the god of the underworld. Mars/Nergal makes its debut in literature in Babylon's Gilgamesh epic (18th century BC), which features a famous version of the flood story in the Bible. Nergal tears down the doorposts of the Upper Ocean to immerse the Earth (an interesting parallel to Mars' own dramatically flooded past). He also turns Gilgamesh's buddy Enkidu into a ghost.

The Babylonians get credit for inventing the zodiac and astrology, and developing an advanced system of mathematics to compute the times of various astronomical phenomena. They composed a sacred week of seven

Stele of Abad-etir. Symbols of Shamash, Sin and Nergal, 9th C BC

days, the 'Seven Pillars of Wisdom' according to Hebrew mystics, and assigned each day to one of the seven prominent 'planets'. Nergal got Tuesday, and along with his underworld tasks he took charge of growth and vegetation—the roots of a plant belong to his underground realm. The Greeks and Romans inherited the Babylonian sacred week, and so have themselves to look too closely at the myths. In our time, the great exceptions are Giorgio de Santillana and Hertha von Dechend, two formidable scholars whose book *Hamlet's Mill* presents startling evidence of astronomical themes in myths from around the globe. The great theme is the precession of the equinoxes and the passing of world ages, when an old heaven

Planet	Influence	Babylonian	Greek	English	
Sun	Illumination	Samas	Helios	Sunday	
Moon	Enchantment	Sin	Selene	Monday	(Moon's day)
Mars	Growth	Nergal	Ares	Tuesday	(Tiw's day)
Mercury	Wisdom	Nabu	Hermes	Wednesday	(Wotan's day)
Jupiter	Law	Marduk	Zeus	Thursday	(Thor's day)
Venus	Love	Ishtar	Aphrodite	Friday	(Freya's Day)
Saturn	Peace	Ninib	Kronos	Saturday	(Saturn's Day)

Note that Mars' Latin name, still lingers on in the Italian (martedi), French (mardi) and Spanish (martes)

we. Note that while English has kept the names of the Roman gods for the planets, it borrowed the day names from their Germanic cousins. To anyone who has given much time to studying the myths and sacred legends of our world, it is abundantly clear that Mars and the rest of the planets are present everywhere, in one disguise or another. Unfortunately, scholars of myth and religion are hardly ever knowledgeable or even interested in celestial connections, while historians of astronomy rarely trouble dies and a new one is born; the planets, by way of the gods and heroes that represent them, are the protagonists. As a basic example of what they call the 'great code' they quote the ancient writer Lucian of Samosata, who thought that the story of Ares and Aphrodite caught in bed (*see* below) was just a silly distortion of some older story, one representing a conjunction of Mars and Venus in the Pleiades (often thought of as a net in the heavens, like the one Hephaestus used). Stories that begin as astronomical myth eventually degenerate, over the ages, into fairy tales.

In de Santillana and von Dechend's writings, Mars emerges as a shadowy and significant figure in surprising places. Befitting his destructive nature, he seems to have a major role in bringing about the end of ages. In this role his disguises include not only the Mesopotamian Nergal, who knocked down the doorposts and caused the flood, but also the long-haired Japanese hero Susanowo, the 'Brave-Swift-Impetuous Male' of the epic *Nihongi*, who pulled his own house down around him—and his Biblical counterpart Samson, who pulled down the house of the Philistines.

Whatever astronomical lore may exist in the Bible is hidden pretty well. Mars—if he isn't Samson—does not appear once in Old Testament or New, at least not in any form that scholars would generally agree on. In the Apocryphal books, however, a few references do shine through the murk and confusion. In the *Secrets of Enoch*, for example, an angel conducts Enoch through the seven heavens; in the fifth, Martian, sphere, he meets a host of giant soldiers called the Grigori. These were followers of Lucifer who rebelled against God and were exiled here, where they bewail their eternal fate night and day.

Rabbi Joel C. Dobin, who studies ancient Hebrew astrology, believes that Mars, or Madim, is hiding in Isaiah 63: 1–6:

> 'he who is born under Mars will be a shedder of blood'. But besides a soldier, this could mean a surgeon, a butcher or a ritual circumciser.

'Who is this that cometh from Edom, with crimsoned garments of Bozrah? He that is glorious in his apparel, stately in the greatness of his strength.

'I speak in victory, mighty to save.'

Wherefore is thine apparel red, and thy garment like he that treadeth in the wine-vat?

'Yea, I trod them in mine anger and trampled them in my fury...'

The *Talmud*, which contains some astrological commentary, notes that 'he who is born under Mars will be a shedder of blood'. Besides a soldier, this could mean a surgeon, a butcher or a ritual circumciser—or even a judge, who inflicts punishments. ●

A Gangster on Mount Olympus

The ancient Greeks named the five restless stars in the heavens the *planetes* or 'wanderers', but for them, like the Babylonians, Mars' uncanny comings and goings, its fadings and brightenings were evil omens—signs of blood and fire, war and lust. The Greek name for Mars, Ares, translates as derring-do or battle frenzy, and the story

> The Greek name for Mars, Ares, translates as derring-do or battle frenzy

goes that this war god came from barbarian Thrace, where people were said to fight simply for the love of fighting. And at least from the time of Homer, the Greeks viewed Ares with a moue of distaste: in the Iliad, he is a whining bully, beaten in one-to-one combat by the goddess Athena (a dame, yet!) who pokes him in the belly with a spear. Bellowing, Ares scurries up Olympus to tattle to their father Zeus. But Zeus brushes his complaint aside: 'You are the most hateful to me of all the gods who hold Olympus; forever strife is dear to you and wars and slaughter.' In Athens, Ares was best known for being the defendant in the first murder trial, on a hill still called the Areopagos, where all subsequent homicide cases were heard. In fact, the Greeks disliked the god so much that they often called his planet Hercules, or sometimes *pyreos*, 'the fiery'.

The only deity who loved Ares besides his own sister Eris (little Miss 'Rumour that stirs up strife') was Aphrodite, the goddess of love, who doted on the big thug. When her husband Hephaestus, the god of fire and metalworking, eventually got wind of their affair, he secretly went to his forge and made a gossamer net of unbreakable metal, which he draped around the matrimonial bed before going off on a divine business trip. In the morning, he raced back to snag Ares and Aphrodite in the net and summoned all the other gods to come and witness his dishonour. The goddesses all stayed home because of modesty, but all the males laughed at the naked and embarassed lovers struggling in the net, and joked that they would gladly bear such a humiliation to be in Aphrodite's arms.

Aphrodite gave Ares two sons, Phobos (Fear) and Deimos (Terror), who must have been a headache to toilet train but grew up to find useful careers harnessing their father's war

Borghese Aries 420 BC by Alkamenes

> **Ares is blind, and with unseeing eyes Set in a swine's face stirs up all to evil.**
> *Sophocles, a fragment*

chariot. The gods of Love and War also produced a daughter called Harmonia, for the kind and affectionate spirit that should spring up in communities after conflicts have passed.

For all their stories, the Greeks were the first to drag cosmology out of the sphere of myth, and to seek out rational theories to explain what they observed. The first Greek astronomers took it for granted that the Earth was the centre of the universe, and that the planets moved about it in perfect circles—at least until Aristarchus of Samos (310–230 BC). To account for the retrograde motions and fluctuations in brightness of the planets, most notably Mars, Aristarchus took a daring leap of imagination and declared that the Earth was a mere planet among planets, and that all planets circled the sun. This was simply too much for the greatest mathematicians and astronomers of the day; who judged the very idea of a non-homocentric universe as outlandish. Instead, the Earth-centred views of other astronomers such as Apollonius, Hipparchus and Claudius Ptolemy prevailed. Ptolemy had the last word, in his *Almagest*, 'the Great Book' (2nd century AD), which offered the elaborate theory of epicycles—smaller spheres set into the great spheres of the planets—to account for every observable movement in the sky. His explanation became scripture until Copernicus. Hand in hand with Greek science went an increasing interest in that Babylonian import—astrology. Ptolemy was a master of it, and he remains the patron and mentor of Western astrologers today. Astrology, in fact, provided the necessity for his 'scientific' work; the ability to predict the motions of the planets with precision made it easier for astrologers—and physicians—to do their jobs right.

One of the inspirations for Aristarchus' precocious theory came from the mystic mathematician Pythagoras, another native of Samos. In the 5th century BC, Pythagoras had suggested that all heavenly bodies circled a central fire, and described their proportional circles as the music of the

De A. Theuet, Liure II.
PYTHAGORAS PHILOSOPHE
Grec. Chap. 25.

spheres—only by observing the motions of the heavens, he declared, could we understand the harmonies of the universe. In the 17th century, Kepler, after discovering the secrets of Mars' orbit, updated Pythagoras' music of the spheres in his third 'harmonic' law of planetary motion (1619); in 1980, one of the mementoes of Earth packed aboard Voyager for its journey out of our solar system was a computer-generated recording of Pythagoras' and Kepler's music of the spheres. But then, it was back in the 4th century BC, when Greek philosopher Metrododrus iterated the reason why we would ever bother sending off a Voyager in the first place: 'To consider the Earth as the only populated world in infinite space is as absurd as to assert that in an entire field of millet, only one grain will grow.'

Rome's Red Grandpappy

It was the Latin tribes of central Italy who made Mars a star, so to speak, and gave it the most-favoured-planet status that it has kept to this day. The name **Mars** (originally Mavors, or Mamers) actually comes from the Oscans, a related Italian people, but Rome's policy throughout its history was to bribe or capture their enemies' gods and get them to change sides. The Mars that the Romans bagged off the Oscans was, like Babylonian Nergal, a vegetation god, in charge of the vines, corn, and pastures. He too had a rendezvous with death; in the month that took his name, March, the god became a scapegoat, the personification of the old dead year, with all of its bad luck heaped on his head—which the Romans, with their wry sense of humour, sent back to the Oscans. On the principle of 'The king is dead. Long live the king!' a fresh young strong Mars would at once take the place of the old one, ready to battle crop demons. Later on, when the Romans decided to hammer their ploughshares into swords, Mars didn't miss a beat, but took his place in the front of the cohorts as generalissimo.

In Roman legend, Mars had played an important role in the founding of the state—by raping a maiden named Rhea Silvia. This girl had been forced into Vestal Virginity by her father, the king, because of a prophesy that his own grandsons would overthrow him. Rhea Silvia secretly gave birth to twins, **Romulus and Remus**, who were set

> The name Mars comes from the Oscans but Rome's policy throughout its history was to bribe or capture their enemies' gods and get them to change sides

adrift in a basket on the Tiber, rescued and suckled by a she-wolf. When they grew up, Papa Mars appeared to them and told them to found a city where they had been raised. As if to show whose boys they really were, they soon set to quarrelling; Remus was teasing Romulus by jumping over his new-built city wall, so Romulus did the Martian thing and killed him. The Roman legends know the exact year: 753 BC.

Early Rome came under the strong influence of the more powerful and civilized **Etruscans**, but the Romans never forgot that Mars had been their top god before the Etruscan sophistos replaced him with Jupiter. Even in Imperial times, a Roman general or emperor leading a triumphal parade up the Sacred Way would stain his face red, in honour of Mars. Originally, as we have seen, he was Mars, as sacred king; later rulers in their triumphs would have a slave on their chariots to whisper continuously to them: 'Remember that you are a man!'

Rome's chief priest, the **Pontifex Maximus** (the title inherited by today's popes) was closely identified with Mars, and he kept the god's sacred spears and figure-eight shields (*ancilia*) in the **Regia**, the oldest temple in the Forum; the spears would rattle whenever Rome was in danger, and when war broke out, the Consul would seize one and shout: 'Mars Vigla!' ('Awake Mars!'). The god's shields were guarded by the **Salians**, the twelve leaping priests of Mars, who each March would transport the *ancilia* in a procession across Rome, hopping, dancing, singing and banging the shields to commemorate the beginning of the war season (but also, it has been suggested, to scare away crop blight). In Imperial times, Mars regained some of his old status when Augustus chose him as his patron while avenging Caesar's assassination; the ruins of Augustus' **Temple of Mars Ultor** (Avenging Mars) still stand near the Forum.

Under Roman rule, worship of

Roman staue of Mars found in Seville.

Mars and the other gods spread throughout the empire. Roman versions usually attached themselves to local deities. In Britain votive inscriptions have turned up with dedications to: Mars Lenus (in Caerwent, Gwent), Mars Rigonemetis (in Lincolnshire), Mars Loucetis ('the Brilliant', in Bath), Mars Toutatis (in Herefordshire) ...among many others. Many stone reliefs on these show geese; perhaps they were sacrificed to the god.

Meanwhile, in China....

Chinese astronomy/astrology is as old as Babylonian (*c.* 2600 BC, and probably much earlier) and its star gazers were just as dumbfounded by Mars' loopy orbit and dimmings and brightenings. In fact, the Chinese name for Mars, **Huǒxīng**, literally means 'fire star': they linked its red colour not with blood but Fire (Huǒ) as one of the symbols of the 'Five Elements'. The other four elements are Wood, ruled by Jupiter; Earth (i.e. soil) ruled by Saturn; Metal, ruled by Venus; and Water, ruled by Mercury.

A major concern of Chinese astronomy was determining when the sun, moon, and planets would line up in the heavens. Over the centuries clocks and calendars were improved, and chains of observatories were set up for the purpose of predicting conjunctions. A study of ancient ephemerides by Kevin Pang of the Jet Propulsion Laboratory and John Bangert of the U. S. Naval Observatory has pinpointed such a conjunction in 1953 BC, which may have provided the starting date for the Chinese calendar.

From the time of the Han dynasty (beginning in 220 BC), other conjunctions were noted as portents of great events—that is, as long as the planets and portentous events happened to coincide. If not, as Y. L. Huang has shown in *A Study on Five-Planet Conjunctions in Chinese History*, historians working for the ruling classes had no qualms at all about concocting the required astronomical event. Most recorded conjunctions never happened at all, while many real, observable events over the centuries were pointedly ignored as irrelevant to Chinese affairs. A favourite excuse for changes in the highest ranks was **Mars in Scorpio** (see p.67), the time when 'ministers resign and the emperor passes away.' Huang found 23 ominous references in Chinese history to Mars in Scorpio, of which 17 were fabricated, while some 40 actual occasions passed by unmentioned in any chronicle. For instance, one such phoney event in 7 BC explained the death of Han prime minister Chai Fangchin, who was 'granted the privilege' of committing suicide by the Emperor Ch'eng, to take on his shoulders all the weight of the unfortunate alignments in the heavens (as in Rome, Mars demands its scapegoats). According to the ancient ephemerides, however, Mars was blameless; it was nowhere near Scorpio at the time. The emperor just wanted to bump off Chai Fangchin.

Fake or real, the Chinese have traditionally viewed a conjunction of the five planets as the best of all omens, one that will 'amass the world...the virtuous will receive celebration and the great men who change reigns will control the four quarters'. The West, however, has often thought the contrary; since the ancient Greeks, astrologers often held that major conjunctions heralded disasters or even the end of the world, a belief that seemed to be confirmed when a conjunction of five planets in the constellation Aquarius in 1345 was followed by the Black Death, and the loss of a third of Europe's population. Chances are you will be around to see who's got it right, the Chinese or the Western doomsayers, at the next great conjunction, 5 May 2000, the celestial event ushering in the new millennium, when the five planets, sun and moon will all meet in the constellation Aries. (Because of the presence of the Sun, you'll only see this at sunrise and sunset. Don't expect all the planets to be right on top of each other—having them all together in one constellation is as close as they ever get.) ●

The First Prussian

Among the Germanic peoples, Mars took the disguise of **Tyr**, or Tîwaz (or Tiw or Tig, among other variants), whose special day became our Tuesday. With the Saxons, Tyr was known as Saxnot, the chief god who gave his name to the tribe, and hence also to Saxony in Germany, and Wessex, Essex and Sussex in England.

The nordic Mars was a one-handed god.

The object of worship in his sacred groves, you can bet Tyr was as warlike a god as Ares or Roman Mars; archaeologists have dug up spears, helmets and swords carved with the runes of his name from Iceland to northern Italy. Teutonic warriors swore their oaths by him before battles, and when cornered they might promise Tyr the sacrifice of the entire enemy force should they come out on top. The Roman historian Tacitus records an especially hard-fought battle between two tribes called the Hermundari and the Chatti, where the berserk victors fulfilled their oath by not only slaughtering all their prisoners, but destroying all the booty too, even the gold and weapons, in the god's honour.

The nordic Mars was a one-handed god. He lost the other one ridding the world of a great menace: the wolf Fenrir, son of Loki and enemy of all the gods. By magic, it was learned that this monster could only be bound by a rope made from intangible things—the roots of a mountain, the noise of a moving cat, the breath of a fish, etc. And by magic, such a rope was obtained; even so one of the gods had to place a hand in the wolf's mouth to be allowed to slip it on, and only Tyr was brave enough to do it.

Tyr wasn't only interested in battle. Unlike the Marses of other mythologies, he seems to have been a friend to man at times, who presided over matters of law and justice. He was especially associated with the *thing*, the tribal assembly common among northern peoples. In a time when public controversies and legal differences were often decided in a trial by combat, no assembly would have been complete without him.

For reasons not at all clear, by late Roman times Tyr was being supplanted as chief war god by his neighbour in the calendar, Wednesday's Wodan, or Odin. Just the same, runic inscriptions to the old god are found well into the Middle Ages. Charlemagne spent much of his career raiding the still-pagan Saxons and chopping down Saxnot's sacred groves, but a monkish chronicle reports the recalcitrant Teutons setting up a column (perhaps a tree trunk) 'in honour of Mars' after a victory over their Thuringian neighbours as late as 970.

Astrology's Middle-Aged Bad Boy

After the great achievements of the Hellenistic scientists, Western speculative thought settled in for a nap that would last over a thousand years. The Roman world cared little for science or philosophy, and while everyone knew that our destinies were controlled by the stars, even astrology was in the doldrums. Astrological fortune-telling did enjoy a great vogue in the early Roman Empire, but later emperors, seeking to stem the tide of irrationality and mysticism that was rapidly turning Roman brains to jelly, outlawed the making of horoscopes on more than one occasion. Under early

> '..the planet is untrue, unlucky and good for nothing.'
> Michael Scot

Christianity, science wasn't possible at all, and astrology only hung on in folk-beliefs.

Both made their comeback in the Middle Ages, mixed up together the way Ptolemy had left them. Despite official discouragement from the Church, interest in the stars picked up steam all through the period. By the 1300s, princes and even popes had their resident star-gazers, and this century also produced the two most astronomically learned poets of all time:

> 'Alas, that ever lovers mote endure
> For love so many a perilous aventure.'

Dante and **Chaucer**. One of Chaucer's prose works is a treatise on the astrolabe, and the *Canterbury Tales* are full of references to the stars and planets. He also wrote a lovely, neglected poem called the 'Complaint of Mars' that might have been inspired by a conjunction seen at twilight. Venus and Mars are lovers in their celestial tryst (Taurus, apparently). 'Gelos Phebus'—the Sun—comes up to spoil their fun, and for valour's sake Mars hangs back to defend her. He gets burned, while Venus escapes to the 'tower of Cilenios'—a learned allusion to Gemini—where she receives some consolation from Mercury.

Any astronomer with the right computer programme could figure out the date when Chaucer saw it happen.

For medieval minds, however, the main attraction of the stars was astrology, and it was in the Middle Ages that Arab, Jewish and Christian astrologers gave their art the form we know today.

It's difficult to find two books about astrology, medieval or modern, that say the same thing, but on the basic nature of Mars all generally agree. Mars is linked with **Aries the ram**, the first house of the Zodiac, and maintained its ancient role as a strongly masculine planet—the Author of War, and also a symbol of energy, change, courage, impetuosity, growth, rage and lack of restraint. The red planet is the fiery one, the agent of heat and dryness; don't expect much rain when it's in the fire signs Aries, Leo and Sagittarius. Mars also signifies blood, and rules over the parts of the body associated with it: the heart, the liver and the veins.

Every planet ruled certain nations, and medieval astrologer Michael Scot adds the somewhat politically incorrect comment that Mars 'lords it over Saracens and other warfaring peoples'. But of course the planet also has a special connection with the island of Britain, home of that renowned 'warrior pirate nation', as Lady Thatcher delicately put it. Britain is under Aries, Mars' own house, and to astrologers John Bull is the archtypical Martial man, a quarrelsome sort who is 'always thirsty', and 'always fighting somebody'; thanks to his Mars-like qualities, he usually comes out on top.

Two bad *hombres* walk the heavenly ways. Worst is slow-moving Saturn, which brings us melancholia and chronic diseases. To astrologers, Mars has always been the 'lesser malific', and its evils are direct and to the point. As one astrologer put it, in the wrong house Mars 'causes death by burns, scalds, fire, cuts, stabs, explosions'. Another is more specific, mentioning 'Abortion, Accidents, Acute Diseases, Assassination, Battle, Bruises, Burns, Bursting of Blood Vessels...' and so on down the alphabet to Suicide, Swords, Syphilis, Venery, Violence, Violent Hemmorhage, War and Wounds...', passing along the way Dog Bites, Fatty Degeneration of the Heart and Passionate Excess. Forget about your healthy diet; just hope this planet doesn't turn up in your horoscope.

The effect Mars has on you does depend a lot on what sign it's in when you're born. Modern astrologers have decided opinions about each, and their books thoughtfully provide examples of celebrity horoscopes to demonstrate. Concerning Mars in Taurus, the watchwords are steadiness and sincerity, along with athletic grace and all-round sexiness, like Shirley MacLaine, Adolf Hitler and Josef Stalin. Mars in Gemini 'often gives expression through the voice' (as with Barbra Streisand or Benito Mussolini), and it is an aspect full of profound intellectual energy (as with Prince Philip, or Dean Martin). Martian power and energy also find an intellectual or artistic outlet in Aquarius, though astrologers caution that a bad aspect can also bring about 'nervous, high-strung and agitated behaviour', and indeed this does seem to be a Martian house full of talented people with whom one would not wish to be stuck in an elevator: Truman Capote, Richard Wagner, Howard Hughes and Charles de Gaulle, to name a few.

It's surprising that Mars in Scorpio does not mean anything special to today's astrologers. To their ancient counterparts, this position would have

> Mars in Taurus, the watchwords are steadiness and sincerity, along with athletic grace and all-round sexiness, like Shirley MacLaine, Adolf Hitler and Josef Stalin

been serious indeed, for Mars has a rival in the skies, and Scorpio is its home: Antares—'anti-Ares'—the big red star that marks the Scorpion's heart, was first mentioned under that name by Ptolemy. It is indeed often confused with Mars by amateur observers. In classical times Scorpio was Martis Sidus, the astrological 'home' of the planet, in contrast to neighbouring Libra, where Venus dominated.

> nothing that grows is more Martian than a raspberry or a hot pepper

Mars sent its influences down to the growing things of our world, and in the ancient lore of the herbalists its manifestations are often only too obvious. Mars ruled over many plants with conspicuously red flowers and fruits. Because of its two-year path around the zodiac, the planet was also in charge of most biennials. Cacti and other desert plants took their characters from Mars, along with anything else that had thorns, befitting the planet's prickly nature. Fruits with an astringent or piquant taste can be Mars' too—nothing that grows is more Martian than a raspberry or a hot pepper. As with astrology, no two sources would agree. The famous herbalist Culpeper would quibble about the raspberries, but he attributes great Martian virtue to radishes and especially garlic, for its properties of cleansing the blood. Despite all the bodily ills associated with the 'lesser malific', herbs ruled by Mars can help by burning up fevers, and cleaning wastes and poisons out of the body.

Mars does have a good side. 'Mars gives man bravery and nobleminded-ness', as the early Islamic encyclopaedists of Basra put it, and on the whole Muslim scholars of the early and medieval periods were inclined towards a more balanced view of the planet. If you've ever wondered where the phrase 'seventh heaven' came from, credit the Muslim sages who translated and reinterpreted the star lore of antiquity. The seven heavens were the spheres of the seven planets, counting the Sun and Moon. Add the heaven of the Zodiac, and the heaven of the Fixed Signs, and you've got nine in all, forming the complete Muslim cosmos. Mars (*Al-Qahira*), fits in right in the middle, the fifth heaven, and as such served as a 'balance point', midway between us and the Creator beyond the stars.

Dante too did his best to be fair. Mars like all the planets has its sphere in the *Paradiso*, and the poet's warlike ancestor Cacciaguida leads him through it. Here, under the vision of a great cross, they meet the souls of those who put their Martial talents at the service of religion: Charlemagne and Roland, Robert Guiscard, Godfrey of Bouillon, and the Maccabees of ancient Israel. But it was only natural for Dante to show some sympathy for the '*stella forte*' that guided the destinies of his own beloved Florence.

Florentine Intermezzo

One would not suspect that the cradle of the Renaissance would have a special relationship with Mars. But the city was founded under its sign by Julius Caesar in the spring of 59 BC, and its oldest symbol was an equestrian statue of Mars, the **Marzocco**, which stood on the Ponte Vecchio until a flood washed it away in the 14th century (a replacement Marzocco was sculpted by Donatello in the form of a lion). In 1215, a heinous murder at the foot of the original Marzocco began the war between the city's medieval parties, the Guelphs and Ghibellines; Martian influences were held primarily responsible for the incessant strife. But Mars was also ruler of springtime, flowery growth sign of Aries; Florence's other symbol was the lily, and its most important holiday the New Year, coinciding with the

Donatello's Marzocco

Annunciation, in March, when Tuscany starts to blossom.

Cosimo de' Medici, Florence's biggest banker and political godfather, was the great patron of **Marsilio Ficino** (d. 1499), the founder of Renaissance Neo-Platonism. Ficino was the first to believe that Christianity should be regarded in the context of older religions; this led to a keen interest in hermetic writings, magic and astrology. For Ficino (and for his follower, Giordano Bruno) Venus, Sol and Jupiter were nothing but beneficial, while Mars and Saturn took their customary role as celestial bogeys; Mars is associated with ferocity, rigidity, truculence, and worse.

Ficino had a direct influence on the creation of some of the most beloved paintings in the West: as chief magus and leader of the Medici-sponsored Platonic Academy, he provided the subjects for the sublime, enigmatic mythologies that Sandro Botticelli painted for the Medici villas. The trio of Sun, Venus and Jupiter appear in the most celebrated of these, the *Primavera*. Mars wouldn't have felt very welcome in such company, but Florence's old gangland patron does make an appearance in *Mars and Venus*, now in London's National Gallery; here, in the height of his allegorical prime, a handsome young Mars naps peacefully, the spirit of War tempered, as he is watched over by a fully clothed, wistful Venus with the features of a Botticelli Virgin. Mars and Venus had been a popular subject for the ancient Roman wall painters, and would become so again in the soft-porn saloon art of the Victorian era. But in the hands of Botticelli and his mystic circle, there are levels of meaning we may never recapture.

The pairing of Mars and Venus lived on through the art of the late Renaissance and beyond, though the

Mars and Venus, Sandro Botticelli

lovely and fragile magical synthesis of Botticelli's time could not. In 1637, Rubens painted his *Horrors of War* (Pitti Palace, Florence); here Mars, led on by the Furies, plunges grimly into a nightmare of gore and destruction while a voluptuous Venus despairs over her inability to hold him back. 1637, not coincidentally, was one of the darkest hours of the Thirty Years' War, the bloodiest conflict Europe had yet seen. The sweet dreams of the Renaissance were definitely over.

A Prisoner at Last

I bring to Your Majesty a noble prisoner...Hitherto, no one had more completely got the better of human inventions; in vain did astronomers prepare everything for battle...Mars, making game of their efforts, destroyed their machines and ruined their experiments; unperturbed, he took refuge in the impenetrable secrecy of his empire...

Johannes Kepler, Commentaries on the Motions of Mars, 1609

After hibernating for over a millennium, the history of astronomy suddenly recommenced in 1543 with the publication of **Nicolaus Copernicus'** *De Revolutionibus Orbium Caelestium*. With one fell swoop, the Polish canon of Frauenburg cathedral replaced the Earth with the sun in the centre of the solar system. This neatly explained a good part of Mars' irascible orbit (the Earth's own orbit around the sun, inside the orbit of Mars, is what makes it appear to move backwards), although like Aristarchus before him, Copernicus thought the orbits must be perfect circles, which led to complicated manoeuvres on his part to explain some of the observable motions.

Copernicus' ideas were not everyone's cup of tea. His fellow churchmen condemned them, as did a number of astronomers, including the quarrelsome but great Dane **Tycho Brahe**. Tycho's considerable contribution to Mars was his faithful and impeccably accurate recording of planetary positions over twenty years (also, no book that mentions Tycho fails to note that he wore a false nose of gold, silver and wax, after losing the original in a duel, so ours won't either). In 1600, he hired an assistant, **Johannes Kepler** (1571-1630), and assigned him the task of explaining the orbit of Mars. The cocky young Kepler thought it would take a week. Instead his 'war with Mars' lasted nine years.

Kepler had the advantage of being a Copernican, and had a lucky break when he inherited the records of the jealous Tycho after his death in 1601. Still, he laboured incessantly, and had to take a year off because the quest was ruining his health, before the great breakthrough came like a flash of light. He realized that Mars' orbit describes an ellipse, with the sun at one focus, and that all the other planets do the same—Kepler's first law of planetary motion (1609) published in *Commentaries on the Motion of Mars*. In 1687, in *Principia*, Isaac Newton would use Kepler's discoveries to elaborate his principle of universal gravitation.

In 1609, just as Kepler published his great opus, another Copernican, **Galileo Galilei**, professor at Padua, constructed his first telescopes and discovered four moons around Jupiter. Kepler congratulated him, and added 'I long for a telescope, to anticipate you, if possible, in discovering two around Mars (as proportion seems to require)...'. A Pythagorean sense of order led Kepler to seek a more simple and elegant explanation for planetary motions than the old Rube Goldberg/Heath Robinson contraption of Ptolemy. It led him to his lucky guess about the moons: Venus and Mercury have none, Earth one and Jupiter four (observed at that time)—therefore Mars should have two.

Someone you'd never suspect—Jonathan Swift—was either a keen student of physics or a very lucky guesser. In *Gulliver's Travels* (1726), Part III, Chapter 3, he tells of the astronomers of the fictional island Laputa, who have likewise

Gulliver sees Laputa

... discovered two lesser stars, or satellites, which revolve about Mars, whereof the innermost is distant from the center of the primary planet exactly three of his diameters, and the outermost five; the former revolves in the space of ten hours, and the latter in twenty-one and a half; so that the squares of their periodical times are very near in the same proportion with the cubes of their distance from the centre of Mars, which evidently shows them to be governed by the same law of gravitation that influences the other heavenly bodies.

It would take another 150 years before Mars' satellites were first seen by Hall; what's more, Swift was spot on in his descriptions of the peculiar orbits of Phobos (0.32 of a day) and Deimos (1.26 of a day) as well as their distances from Mars (9379 km and 23,459 km, respectively). Did the great Irish writer have inside information?

In 1771, astronomer **William Herschel** took a good look at Mars and, noting that its frozen poles and seasons appeared to be much like ours, declared that it was so like Earth that it could probably support life, and that its inhabitants 'probably enjoy a situation in many respects similar to ours.' It was the first time a modern scientist suggested such a thing, but it certainly wasn't the last. ●

Made in Italy—the Martian Canals

The story of the canals is a genuine parable. It begins in 1858 at the Collegio Romano, the mouldering old Jesuit university a few blocks away from the Pantheon in Rome. Rome was entering its last decade of papal rule then, and the Church slumbered fitfully in its dotage; the Inquisition was still in business, and Galileo and Copernicus still on the blacklist. Down in the cellars of the Collegio, a drunken porter was selling off Italy's greatest hoard of Renaissance documents and manuscripts for scrap paper to buy wine (they caught him years later when a scholar bought some butter in the market and found it wrapped in a letter from Christopher Columbus).

Nevertheless, a few men in Rome were still wide awake. One of these was the Jesuits' chief astronomer, **Pietro Angelo Secchi**. 1858 was a Martian perihelic opposition year, and Secchi took the opportunity to do some extended observations from the Collegio's observatory. He had a good look at the 'hourglass sea' (now Syrtis Major), and it seemed to him to resemble the narrow stretch of the Atlantic between Africa and the Americas. Like most astronomers of his time Secchi was convinced that the dark patches were really water, and consequently called this one the *canale*, or channel. One of the great Martian myths was born.

At the same time, another worthy Italian astronomer was just starting out. **Giovanni Virginio Schiaparelli** became the head of the Brera Observatory in Milan in the 1860s. In a long and distinguished career, Schiaparelli did important work on double stars, and on Venus and Mercury, discovering the periods of rotation for both. Nevertheless, it is for his studies of Mars that he is best known; he drew maps of the planet that were excellent (for the time), and designed fancy classical names for

Schiaparelli

many of its surface features that are still used today. He also was probably the first person to observe the Martian dust storms.

Schiaparelli had an excellent eye for details, but he also happened to be colour blind. This seems to have actually helped him with his observations; the little handicap made it possible to pick out nuances of shading that other men couldn't see. Ironically though, it also helped him to see things that weren't really there. In his long and careful observations of Mars, Schiaparelli discerned a number of long, straight streaks across the planet's surface, and following Secchi he called them *canali*. These were very faint and difficult to observe; the astronomer noticed that the planet had to be turned just the right way for features on any one part of it to become manifest. The more he looked the more he saw, and over the years he built up a network of intersecting *canali* that girdled the planet. By the 1870s Schiaparelli was talking about 'gemination' or twinning; many of the *canali* seemed to be made up of two fine parallel lines. At about the same time, the first spectrographic observations revealed the presence of water vapour near the poles. There was water on Mars, and for anyone with just a little imagination, the *canali* had become canals.

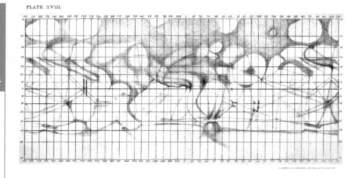

Schiaparelli's map of Mars 1881–82

New Earth, New Mars

At that time, you'll remember, our own planet was changing almost as fast as Mars. Mass-circulation newspapers, the telegraph and later the telephone, mass education and the dissemination of popular books on scientific subjects were creating a new mind-set, a new sensibility. Millions of people around the world were turning their attention to the heavens, and they were more inclined than ever before to wonder, and to cast their imaginings in the terms that science presented. Not surprisingly, there was as much room for science fiction as for science fact. While Schiaparelli was star-gazing, **Jules Verne** was already writing his first novels about trips up to the Moon and inside the Earth.

As much as anything, it was Mars and its 'canals' that put the spark to this explosive mixture. Many began to speculate that the planet might possibly be inhabited, and even the wildest of these speculations found a hearing. One American declared that part of Schiaparelli's canal network clearly spelled out the Hebrew word for 'Almighty'—the Martians were already signaling us, proclaiming their faith to the neighbouring planets. The papers noted that this gentleman was a confirmed agnostic, and therefore unmotivated by prejudice. Anything was possible.

Astronomers around the globe sought out Schiaparelli's canals for themselves. Many found them without much trouble, and even added new ones. Others found them difficult, and wished they had better conditions for viewing and bigger telescopes. It wasn't just canals; Mars was pouring out water everywhere. One university's observatory reported finding over forty small lakes on the planet, while another observer claimed that an entire continent had been suddenly flooded; it 'reappeared' two years later. Nevertheless, it was a Boston Brahmin named **Percival Lowell** who really put Mars and its canals on the map—figuratively and literally. The man who became the true apostle and patron saint of all the Little Green Men from space was the scion of that famous family who 'spoke only to God' as the old saw had it, and who gave their name to Lowell, Massachusetts. A charismatic dilettante aristocrat, he spent some years writing now-forgotten books about the cultures of the exotic Orient, among other subjects, before finally deciding to devote himself to astronomy in 1893. With his wealth, connections and gift of blarney, Lowell was able to hijack a Harvard University expedition to Flagstaff, Arizona, where the clear desert air made for perfect viewing conditions.

Within a few months, Lowell was posing as the Renowned Expert. As one astronomer put it: 'One would think he was the first man to use a telescope on Mars.' Though at first he had a hard time with the canals, enthusiasm soon made up for the eye's shortcomings. Lowell's drawings showed the canal network as geometrically precise, and obviously the work of intelligent beings. He evolved an elaborate, and, on the face of it, quite logical theory to explain them, in which the Martians had to build the canals as a matter of survival, drawing down water from the poles to nourish their dessicated planet. Soon, the idea grew into a theory of the life-cycle of planets, where each gradually aged and withered under the advancing deserts. The Moon was already dead, and Mars was dying—though its advanced civilization knew no war, as everyone's maximum effort was required for the upkeep of the canals. Serious scientific journals wouldn't publish his findings, but the newspapers and popular magazines did— Boston's own *Atlantic Monthly* could hardly refuse. Lowell gave lectures around America and Europe, proclaiming his proofs of advanced life on Mars. He got his first book, *Mars*, out in 1895, and it was a sensation. Everybody was talking about Martians, and Percival Lowell had made himself the first celebrity astronomer.

> He evolved an elaborate theory to explain them, in which the Martians had to build the canals as a matter of survival, drawing down water from the poles to nourish their dessicated planet

Among the scientists at least, there was still a solid bedrock of scepticism about the canals. In particular Alfred Russell Wallace, co-founder of the theory of evolution, took pains to refute the Lowellian moonshine, assembling the evidence and finally screaming off the page in exasperation, 'the planet is UNINHABITABLE!' Not many people were listening. In 1903, two British astronomers named **Evans** and **Maunder** conducted an experiment in which they hung up a sketch of a planet that looked like Mars (*sans* canals) at the front of a classroom, and asked the schoolboys to copy it. They found that when an object with light and dark patches could not be seen clearly, there was a decided natural tendency for the eye to connect the darker patches with dark lines or strips—as Schiaparelli had done. The further back in the classroom the students sat, the more likely they were to do this.

Lowell and his disciples dismissed what they called the 'small boy theory' with noisy contempt. Meanwhile, they were busy expanding their theories to the rest of the solar system, proclaiming the existence of canals and similar markings on Mercury, Venus, and even one of the moons of Jupiter. In 1906, Lowell published his definitive book, *Mars and its Canals*, even though his own assistants had conspicuously failed to find any trace of water on Mars in their spectrographic experiments. As astronomers around the world looked at Mars through new and bigger telescopes and found no traces of canals, Lowell claimed that it proved his conviction that only small telescopes were good for looking at planets. In the press, and among scientists, the canal wars raged on; the best astronomers of the day lined up on one side or the other, or declared themselves canal 'agnostics'. Some artfully took both sides, like old Schiaparelli, who wrote letters alternately supporting and denying the canal theory until his death in 1910.

Lowell's favourite portrait

Lowell himself died in 1916, a fervent evangelist until the end. Though support among scientists for the canals gradually dwindled, the idea had impressed itself so strongly on the popular mind that it poisoned planetary studies for decades; most young astronomers starting out decided to avoid the controversies and study something else. Myths do not die easily. In the 1950s serious scientific books still discussed the canals, and NASA maps and documents mentioned them as late as 1965. The Mariner and Viking photos, of course, put an end to the canals once and for all, but we will never forget them. As we said, the canals are a parable, and full of weighty matter. The story of how science could raise up an outrageous mountebank like Percival Lowell is only one of its lessons; humanity even at its most earnest can be fooled some of the time, and we should not be so confident as to think that such a thing could not happen again. The greater lesson is a reminder that our eyes and minds may not be as free and autonomous as we think. Lately, science has been telling us with increasing insistence that the eye is not a camera. Only a part of the images we see comes in through the pupil; the mind supplies the rest. When one man saw canals, thousands of others could suddenly see them too.

The War God Awakes

The nice thing about science, of course, is that scientific absurdities eventually collapse of their own weight, no matter how long it may take. As a scientist, Lowell may have a lot to answer for, but his true place in history lies elsewhere. Without his vision of a troubled Mars inhabited by intelligent beings, science fiction and the popular imagination might never have developed the way they did. Lowell put Mars in the centre ring of the planetary circus, and the Red Planet would never be the same again.

The idea was extremely fetching. The immense size of the *canali* predetermined that the Martians were at least as technically advanced as we were, if not more so. Storytellers at once took up the challenge to describe

'An eye like Mars,
to threaten or command.'
Hamlet II, iv

what these Martians might be like and how we could get in touch—in 1880, England's Percy Greg was already there with the first scientific description of a trip to Mars in *Across the Zodiac: The Story of a Wrecked Record*. Greg's Martians not only live in a utopia, but go so far as using telepathy to control wrong thoughts. Other writers suggested that Mars was populated by reincarnated humans (French astronomer Camille Flammarion's *Urania*, 1890), or by advanced, high minded social reformers (L. Edgar Welch's *Politics and Life on Mars*, 1883), or, more mundanely, that it was just like the earth, with a similar development, complete with a Martian Christ (James Cowan's *Daybreak: A Romance of the Old World*).

The first rocket societies were founded at the end of the century, after Russian Konstantin E. Tsiolovsky published his *Exploration of Cosmic Space by Means of Reaction Devices* (1896), with formulas and recipes for a rocket and its fuel. Earthlings fascinated by Mars were starting to think about going to visit—but the Martians got here first.

In **H. G. Wells**' 1897 classic, *The War of the Worlds*, the high-tech, canal building Martians get tough. Once similar to humans, the Martians have used their advanced technology to evolve into arrogant, blood-sucking octopus-like creatures. They have eliminated micro-organisms and diversity from their planet. Worst of all, they never learned to share.

...At most terrestrial men fancied there might be other men upon Mars, perhaps inferior to themselves and ready to welcome a missionary enterprise. Yet across the gulf of space, minds that are to our minds as ours are to those of the beasts that perish, intellects vast and cool and unsympathetic, regarded this earth with envious eyes, and slowly and surely drew their plans against us. And early in the twentieth century came the great disillusionment.

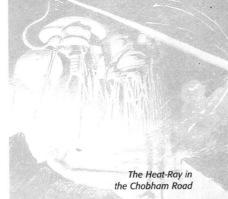

The Heat-Ray in the Chobham Road

Ten enormous machines land in southern England and attack London and march up the Thames valley on long spindly legs (in the original illustrations, the capsules resemble Japanese barbecues on Meccano tripods, with pinchers for Englishman-snatching). Nothing can stop the mighty barbecues; resistance is met by awful alien weapons: a 'Heat-Ray' and toxic Black Smoke. The Martians have it all their own way, until all at once their machines stop and they themselves succumb to Earth's greatest natural defense: germs. The Martians have no immunity.

When *War of the Worlds* was written, many in Europe feared the militarization of a recently unified Germany, and felt at the same time a *fin-de-siècle* uneasiness about the Industrial Revolution, which was proving to be a Faustian bargain at best. Besides the slums and misery of the workers, there was the development of new weapons of destruction; technology, the handmaiden of humanity, was careening out of control, often seeming more witch than servant. Darwin's ideas of natural selection had been around for thirty years, and Wells' ending to his novel wasn't original—Martians had already succumbed to Earth germs in Hugh MacColl's 1889 novel, *Mr Stranger's Sealed Packet*—but Wells provides a grace note of optimism for us.

In the *War of the Worlds*, Mars firmly finds its contemporary role as our essential Other. As it had before, in the age of myth, now every chapter in the story of Mars would reflect Earth's own deep-seated fears and terrors, as well as its longings and dreams, its hopes and desires … among many others.

The War of the Worlds, American pulp cover

A Not-so-Lonely Planet:
Cultural Exchanges in the early 20th Century

Lowell's fantasy of a dying civilization on a dry planet also caught the fancy of American writer Edgar Rice Burroughs, better known as the creator of Tarzan. Burroughs sent an Earthling, Confederate Civil War veteran John Carter, to Lowellian Mars in a series of eleven novels, the *Barsoom Tales*, in which the Red Planet is still habitable, but a mess: the oceans are drying up and the Martian kingdoms are constantly battling each other while trying to ward off environmental catastrophe. Wells' Mars sent us tentacled, goggle-eyed creeps, but Earth turned the other cheek and gave, in return, a frontier-style hero in the classic mould. John Carter fights for truth and justice on Mars; he marries a local girl, has kids, and reaches the upper echelons of Red Planet politics.

Thanks to Lowell and Burroughs, enough Americans sincerely believed in life on Mars that the next logical step, making contact, was pondered in various degrees of silliness. One idea, solemnly endorsed by the *New York Times*, was to construct a proof of the Pythagorean theorum on the Siberian steppes, several miles long, big enough to impress the Martians with our cleverness when they peeped at Earth through their telescopes. Fortunately, an easier solution appeared thanks to a wonderful new tool, radio. Why not establish voice contact with Mars? Transatlantic signals had been picked up successfully in November 1919, and it was reasoned that if the Martians could build such swell canals, they were sure to have super high-powered radio transmitters. It was noted that on 23 August 1920 Mars and Earth would be a mere 34 million miles apart, the closest approach since 1804. Perhaps Mars had been sending messages all along, but Earth lacked the technology to tune in—until now. As any home with a radio had the potential to pick up a Martian broadcast, radio stations across America were requested to go periodically silent on 22–23 August.

It wasn't long before reports came in that Mars was on the air. Even before 22 August, radio operators in Vancouver received mysterious rapping signals; in London, a specially constructed super-radio produced 'harsh notes', which were also heard in Newark, New Jersey. The most notorious incident of all occurred on station WHAS of Louisville, Kentucky on 22 August. By chance, WHAS had just hooked up phones to bring in the world's first live broadcast of military exercises, and the end of the programme was filled with artillery fire. No one had ever heard the like: not surprisingly, many listeners who unknowingly tuned in believed they were listening to a rambunctious message from the planet of the war god.

One earthling did speak with Mars,

> it was reasoned that if the Martians could build such swell canals, they were sure to have super high-powered radio transmitters

as readers of Don Marquis' column in the *New York Sun* soon learned. The news was relayed through archy, a free verse poet reincarnated as a cockroach, who would type his poems at night on Marquis' machine, all in lower case, since being an insect he couldn't manage the shift key. Martians, it turns out, were great admirers of archy and poetry but had some doubts about the rest of us. At one point in the conversation archy asks what the Martians call our planet and, rather shocked by the response, he asks why. Mars explains:

> not long ago one of our prominent
> scientists got a good look at it
> with a new fangled telescope and
> he laughed himself to death crying out
> goofus goofus goofus all the time
> he said from the way it looked
> it couldn't be named anything else
> but goofus–
>
> archy on the radio,
> from archy's life of mehitabel

Learning the Martians' opinion of Earth did not keep Earthlings away for long. The new technologies of the day inspired **Hugo Gernsback**, the 'father of science fiction' (1884–1967), who borrowed the 18th-century tall-tale-spinning character Baron von Münchausen and transported him to Mars in a series of scientific adventures, first published in his own *Electrical Experimenter*, and then in his later magazine, *Amazing Stories*, the first and one of the finest sci-fi pulps. There was plenty of material out there for amazement, with Einstein's theory of relativity kicking around (since 1905) and Clark University Professor Robert Goddard's first experiments with rockets, which he boldly claimed would one day take us to other planets. Goddard later recalled that his career was determined when he was a small boy in Massachusetts; sitting in a cherry tree reading *The War of the Worlds*, he decided that some day he would find a way to go to Mars. ●

War of the Worlds, Part II: All's Wells that ends Welles

On 30 October 1938, when the Earth was once more filled with premonitions of war, 23-year old **Orson Welles and his Mercury Theater** broadcast their notorious radio version of Wells' story in the United States. In the script, adapted by Howard Koch, the Martians alight at a place called Grover's Mill and melt down swathes of New Jersey and New York City with their heat rays. Cleverly faked news bulletins, interrupting a pseudo-performance of the 'Ramon Raquello orchestra', and 'live reports' with realistic sound effects from the invasion scene sowed panic across America—in spite of repeated disclaimers during the show that it was only fiction. Much to his own surprise (he actually thought the story was pretty stupid), Orson Welles demonstrated the brute power of the electronic media to sway the minds of the public. Welles claimed it was only a Halloween prank, but it was an especially ominous one; if an estimated million souls across America genuinely believed that Martians were vaporizing New Jersey, it's hardly surprising that millions of Germans could fall under the spell of Hitler and company, spitting a more plausible nastiness out of their home radios.

In Hollywood, Universal was just finishing the sequel to the popular **Flash Gordon** serials, starring ace swimmer Buster Crabbe, when the Welles furor hit. Immediately, the pro-

> And never was Mars more threatening than under the tyrannical usurper Ming the Merciless, with his evil moustache and long pointy fingernails.

Flash Gordon's trip to Mars, Universal 1938

ducers jigged the script and reshot some scenes, moving the action from the mythical planet Mongo to Mars. The result, *Flash Gordon: Mars Attacks the Earth*, became one of the genuine classics of lowbrow sci-fi. The spaceships may be smoking, spark-spitting from household appliances, and sometimes you can see the strings holding them up, but nevertheless this low-budget serial had a funky Art Deco stylishness that will never be surpassed. And never was Mars more threatening than under the tyrannical usurper Ming the Merciless, with his evil moustache and long pointy fingernails. Mars itself, interestingly, was an attractive and diverse place, with friendly folk like the winged Lion Men and the Clay People to give Flash a hand. As on the Earth in 1938, the only problem was the evil dictator.

Ironically, the more the public believed in Martians, the more careful observers of the planet were ready to throw them out with the canal water. Astronomers in Europe (especially E. M. Antoniadi in France) were producing accounts and sketches of Mars' changing climates, its deserts and vast natural features that had nothing at all to do with canals.

The Postwar Era:
Mars in a Dark Funhouse Mirror

Mars really came into its own after the war; as the cynical, scared, post A-bomb mood of the Cold War ushered in a golden era of **sci-fi pop culture**, our planetary neighbour became the favourite setting for alternative earth scenarios; the stock 'little Green Men' were often enough godless Reds in disguise, or in the case of Warner Brothers' cartoon *Marvin the Martian* (first appearance, 1948) a tricky but lovable little Roman legionnaire equipped with a ray gun. Then there is the whole extraordinarily complex Martian saga of the Justice League in DC comics, with hero J'onn J'onzz and enigmatic shape-changing Green and White Martians who, as episodes go by, appear either as brawny, violent warriors, fire-fearing lizards or gumby ballet dancers.

All of the troubles of the Cold War period were projected by writers and filmmakers onto Earth's nearest neighbour, Mars the scapegoat. In *Rocketship X-M* (1950) the Red Planet suffers a nuclear holocaust; in the classic film *Invaders from Mars*, everyday terrestials fall victim to a villainous Martian mind control scheme; insidious Martian brainwashing also plays a big role in Kurt Vonnegut's *Sirens of Titan*. Anxious hormones from the Eisenhower era, when women began to feel restless confined to the kitchen and nursery, are tangible in the flicks about female Martians (*Devil Girl from Mars*, and the later *Planet of Blood*). In the meantime 'reality' began to imitate fiction; the modern epidemic of UFO sightings began in the 1950s, when an American pilot reported an object resembling a 'flying saucer' (presumably from Mars). Just as everyone once saw the canals, now thousands began to see UFOs.

But who was the enemy really? Ray Bradbury opened up a whole new chapter on Mars and interplanetary relations in *The Martian Chronicles* (1951). The tables have turned: Mars now hosts the wise, just, benign civilization and we humans (well, Americans actually) are the destructive alien invaders, behaving badly and selfishly, destroying another culture for profit before killing off the natives with a home brew of viruses and bacteria. These good Martians play an off camera role in *Red Planet Mars* (1952), a dopey piece of McCarthyism that sets the news of a God-ruled Martian utopia against the cursed Soviets; they also figure a decade later in Robert Heinlein's *Stranger in a Strange Land* (1961), the best of several 'messiah' sci-fi novels of the day. The book looks ahead to the 1990s, when Earth has sold its very last scruples to commerce (who says sci-fi isn't prophetic?). Into this corrupt materialistic world comes a young human raised by the Martians. The hero inherits the lofty moral perspective of his foster parents and their psychic powers; as he learns to cope with his fellow humans, he becomes the centre of a new religion of love that offers the benighted folks of Earth spiritual awareness before the inevitable sad end.

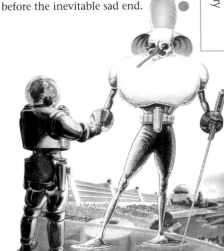

Early Red Successes and Setbacks

The Russian launching of Sputnik on 4 October 1957 launched America's paranoia and competitiveness (Martian attributes, both) into high gear. The USA's fond preconceptions of technological superiority were shattered, and the country felt vulnerable.

In the spirit of the day, the Topps Bubble Gum company came out in 1961 with a series of cards called *Mars Attacks!*, telling the story of an invasion of extremely malicious bulbous-brained Martians, but the cards were soon banned for being too threatening and gory; this was, after all, the time of the Cuban Missile Crisis. (Now that the coast is clear, the cards have been brought to the silver screen in the camp movie of the same name—perhaps the first flick ever to be inspired by gum cards.) Amidst all the troubles, the beginnings of a new school of realistic Mars fiction begins with *The Seedling Stars* by James Blish. To colonize a place like Mars, Blish argued in his stories, either the planet would have to be changed to become more earth-like (he invented the word *terra–form* for the process) or we humans would have to bio-engineer ourselves to fit a different environment.

The Reds of the '50s must have felt a special affinity with the Red Planet. Although they kept their space programme (especially their failures) top secret, it is now known that their first rocket to Mars blasted off in October 1960, but failed to reach Earth orbit; in 1962 two probes were launched, the first failing to reach Earth orbit, while the second, Mars I, actually made it to Mars, but passed silently by, as radio contact with Earth had been lost three months previously. It was only the beginning of an incredible streak of bad luck for the Russians; all in all, they have launched 17 dud or near-dud missions to Mars. Donna Shirley, designer of the Sojourner rover, calls the gremlin that plagues such missions 'the Great Galactic Ghoul'.

Early Mars Tourism:
Postcards from Modern Mariners, 1964-1971

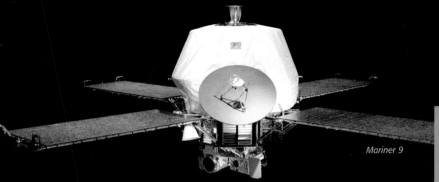

Mariner 9

Age Begins

By the time the next launch window opened, in autumn 1964, the Americans were ready to join in the assault of Mars with **Mariner 3** (which failed to achieve a proper orbit) and the back-up **Mariner 4**, which did. The Russians followed two days later with **Zond 2**, which like their Mars 1 was to suffer from a failure to communicate.

Ninety percent of all our previous assumptions about Mars began to crumble on 14 July 1964, when Mariner 4 popped out its camera and took the first photo of another planet taken by a spacecraft. In all, Mariner sent home 22 blurred black and white pictures, covering just one percent of the Martian surface. And what they showed was craters, nothing but craters—-distressingly enough, Mars looked exactly like the Moon. No water, no vegetation, no canals, just a cloud here, or bit of frost there on a crater wall. The surface was judged to have been dead and unchanged for the past 2 to 4.5 billion years. On Earth, the photos were viewed with keen disappointment, although the pro-canal faction had yet to lose hope—after all, the tiny fraction of Mars photographed was not a big canal area.

The race to Mars was on. Russia blasted off two more failures in 1969, and the USA two more successes: **Mariners** 6 and 7, which reached Mars in July and August, just after Neil Armstrong took the first steps on the Moon. Both craft passed within 3000km of the Red Planet, and had better cameras, which managed to capture some 10 per cent of the surface. Canals were nowhere to be seen, though a few of them turned out to be strings of dark floored craters or canyons. The photos showed signs of erosion (hinting of climatic changes), and polar ice, while Hellas, long believed to be a lofty area, turned out to be a smooth plain. Yet the photography was such that Nix Olympica (since re-named Mons Olympus) looked like a giant crater.

In the spring of 1971, with the space race hot and nasty, the Russians and Americans both launched ambitious missions, each hoping to outshine the other in speed and fact gathering. The US **Mariner 8** failed shortly after lift-off, live on TV, when its guidance system failed and the rocket crashed into the ocean. Two days later the Russians launched **Mars 1**, but through an error relayed to an onboard computer it failed to leave Earth's orbit. With the score still at 0 to 0 for the year, the Russians then successfully launched two enormous rockets, **Mars 2** and **Mars 3** in May; the US followed a few days later with **Mariner 9**; its original mission was quickly adapted to fill in for its lost mate, though left flexible, and open to last-minute decisions.

The three spacecraft raced to Mars, and Mars prepared to welcome them—with an enormous yellow dust storm that completely veiled the planet's surface. Mariner 9 got there first, and went into orbit, becoming the first American craft to circle another planet, and on orders from Earth, shut down its television cameras to wait out the tempest. The hapless Russian orbiters were pre-programmed and could do nothing but send back photos of the dust ball. Each sent down a lander: the first one, as it crashed just north of Hellas, managed to drop a

Mars' Grand Canyon was discovered and named for the spacecraft that found it, the Valles Marineris.

metal Soviet flag, the first man-made object to reach the Martian surface; the second Russian module landed safely, turned on its television cameras, and within a few seconds broke off contact—it may have been blown over in the wind.

This left Mariner 9 the sole survivor of the original five, and after the dust storm subsided, its cameras were switched on again. The spacecraft turned out to be a greater success than any one had hoped for, and before it gave up the ghost on 27 October 1972, it furnished 31 billion bits of information and 7239 high resolution pictures of Mars, Phobos, and Deimos. The new information once again revolutionized previous presumptions about Mars, revealing such an unexpected array of topographical features that in 1973 the International Astronomical Union introduced a whole new set of Mars-specific geographical terms. The volcanoes were finally seen to be volcanoes, not craters, and Mars' Grand Canyon was discovered and named for the spacecraft that found it, the **Valles Marineris**.

In popular culture, news that Mars seemed to lack canals, ancient cities and little green men seemed to induce a hangover to the imagination, along with a certain wistfulness, as in David Bowie's 1972 *The Rise and Fall of Ziggy Stardust*. Much of what had once seemed menacing or alluring was reduced to silliness, as in the films *Santa Claus Conquers the Martians*, *Mars needs Women* or the American TV sitcom *My Favorite Martian*. On the other hand, the somewhat unpromising nature of Mars challenged scientists into theorizing exactly how the planet could realistically be colonized: Carl Sagan's *Cosmic Connection* (1973) was the first to undertake the scientific possibilities of terraforming Mars.

Phildickian Mars: Realities and Paranoias

In the Mars reality hangover, however, the best sci-fi writing changed direction, spawning a new genre known as cyberpunk. Technology, running amok since the time of H. G. Wells, became more sinister than ever, especially in the works of cyberpunk pioneer Philip K. Dick (1928–82). Phildick (as his fans, the 'DickHeads' call him) evoked America in the 21st century as nightmarishly bleak, paranoid and chaotic, where government is a vast conspiracy against its citizens—Vietnam and Watergate seemed to confirm the idea—and every individual's acts are monitored in a techno netherworld. The main characters in this Wasteland revisited are anxious anti-heroes, fighting through the layers of lies and discovering realities that turn out to be fluid, illusionary, multiple; his later stories are said to anticipate Chaos theory.

Mars, in its traditional role of Earth's Other, was one of Phildick's favourite sets in the '60s. In *Martian Time Slip*, the Red Planet is choking with thirst and polluted with schizophrenia, which warps the minds of its inhabitants; in *The Three Stigmata of Palmer Edlritch* Martian colonists are only able to cope with their environment by using hallucinogenic drugs. The story *We Can Remember It for You Wholesale* features a man who purchases a memory of a fake trip to Mars that distorts his real memory; in *Do Androids Dream of Electric Sleep?* (*Bladerunner*) human 'replicants' have

We have the met the enemy and he is us.
Walt Kelly

been created to exploit Mars but try to pass as real humans to take over the Earth. The fact that Phildick, initially published in cheap pulp paperbacks, is increasingly considered one of America's most thought-provoking mainstream writers of the 20th century, speaks for itself; in his lifetime the Japanese wanted to nominate him for the Nobel Prize in literature.

As Leif Erikson and his Vikings were the first Europeans to set foot on the old New World, it seemed appropriate to NASA to lend their name to Earth's first foothold on the next New World. The Viking project to Mars was NASA's most elaborate mission to date; after nine years of research and development, employing experts from around the world to devise instru-

Bicentennial of the United States.

The first Viking entered Mars' orbit in June, and circled for a month, looking for a likely place to roost (the planned July 4 touchdown had to be delayed, as the planned landing site at Cydonia was shown to be far too rugged). This turned out to be the Chryse Planitia and then it was only by luck that the module, the size of a

Mars, invaded by Vikings, tastes Chicken Soup

Viking 2 launch

ments and experiments, identical twin Viking spacecraft, both equipped with orbiters and landers, were launched two weeks apart in August and September 1975. The goal of their mission: search for life, photograph and map the surface of Mars, and test heat shield and parachute technology for landing. The **Viking 1 lander** was designed to alight with a great huzzah on the fourth of July 1976, for the

Volkswagen Beetle, weighing 2633lbs and packed full of delicate instruments, didn't alight on a boulder and fall on its face. The **Viking 2 lander** touched down at Utopia Planitia, some 4000 miles away from its twin, and found itself in a very similar rather dull rocky landscape.

The search for signs of life was the main business of the day. Both landers had compact biology laboratories in their guts. Mechanical arms reached down and scooped up handfuls of Martian soil and rock and stuffed them into their testing chambers, where three different experiments were performed, then repeated with variations to test for signs of life. In the Pyrolitic Release Experiment, the test chamber was filled with carbon dioxide and traces of radioactive Carbon 14. A xenon sun lamp shone

> Both landers had compact biology laboratories in their guts.

on it for five days, and then the Martian soil was heated. In the vapour that arose, organic molecules would be trapped in a gas chromatograph tube and measured for radiation. The count was positive, and suggested some kind of biology might be responsible for the exchange of carbon dioxide from the atmosphere to organic molecules. When repeated, the soil was sterilized with heat, and in another experiment the light was left off. These experiments were negative.

In the Gas Exchange (or 'chicken soup') Experiment, a nutrient and water was added to the soil sample, in an atmosphere of carbon dioxide, laced with helium and krypton. When moistened, the Viking's instruments registered a surprising surge of oxygen from the samples—something that had never happened in any earth samples. Finally, in the Labeled Release Experiment, a moist carbon 14 enhanced nutrient was introduced. At once there was a burst of carbon 14 radioactivity that at first convinced the NASA team that they had hit paydirt: Dr. Norman Horowitz, the designer of the Pyrolitic Release Experiment, said, 'It is not easy to point to a non-biological explanation for the positive results.'

Since then, however, Dr Horowitz and the majority in the scientific community have found their non-biological explanation: exotic chemistry. The apparent positive results derived from Mars' unusual oxidizing rusty soils. The scientists also point out that Viking's Gas Chromatograph/Mass Spectrometer (GCMS), a fancy geological instrument, looked for organic molecules and found none. They also pointed out that life as we know it has, as a prerequisite, liquid water in the sun; nor could it survive the ultraviolet radiation that bombards the surface of Mars. The official NASA statement at the end of the day was that the biological tests were inconclusive.

Some aren't so sure, including Dr. Gil Levin, creator of the Labeled Release Experiment. For one thing, no one has yet been able to reproduce the results of the Viking experiment with exotic chemistry. Also, it appears that the GCMS aren't always reliable. Used in Antarctica, they have failed to note any organic molecules in the soil—which has been shown to be full of life. And if UV radiation has sterilized the Martian surface, argue the scientific minority, couldn't a few hardy microorganisms survive under the surface or in the shadows?

If the search for life was left up in the air, the Viking mission was in other ways a tremendous success. Intended to last only three months, the Vikings sent their last signals to Earth six years later, in 1982, after mapping 97 per cent of the planet's surface and relaying 52,000 photographs of Mars and Phobos. The most famous of these snapshots has become the subject of an enduring cult fixation: Cydonia, the Rocky Horror Picture Face.

Chryse Planitia from the Viking Lander I

The Photo of the first Martian

If one result of the Viking mission was to reduce the odds in favour of life on Mars, another result was to reheat the leftover theory that the planet once hosted an advanced civilization. It began on 25 July 1976, when the Viking 1 orbiter sent back a photo of **Cydonia** showing a hill shaped like a chubby, helmeted face, about a mile in diameter. Viking project scientist Gerald Soffen joked that he was seeing 'the first actual photo of a Martian.' When the photo was published in America, the general opinion was that it looked just like Ted Kennedy. Kennedy's response: 'I guess I should stop eating so many Mars bars.'

The joke didn't stop there. Although planetary geologists were quick to point out that the 'Face on Mars' was a result of natural erosion, and that the play of light shadows in the photo created the illusion, hundreds of other 'experts' (oh where is Percival Lowell now when they need him?) believe the Face is artificial, made by some intelligent life form. There are other mysterious 'artifacts' in the Viking photos of Cydonia to keep the Face company: the so-called City, the Fortress, Cliff, the Tholus, and a 'D&M Pyramid'. Could they be a temple to the Martian moon god? A secret message from the past to drive earthlings crazy? Evidence 'proving' that these features are alien-made include measurements of angles and distances that describe mathematical relationships—especially the D&M Pyramid, which was worked out to a fine degree of precision by Stanley V. McDaniel and published in *The McDaniel Report* in 1994, a scientific-looking account showing that the Face is no accident of nature. McDaniel and others claim that NASA and the US government know that the truth but

'There are probably only six people in this room who know how true this is'

Ronald Reagan, reportedly whispering to Steven Spielberg during a screening of E.T. at the White House in the Summer of 1982

don't want to shock the American public with it; that, they say, is why NASA has yet to conduct a (known) scientific study.

Tapping on this spirit of paranoia, Hollywood came out with *Capricorn One* the very next year (1977). The plot involves the first mannned mission to Mars, which abruptly has to be called off at the last minute because of technical troubles. Rather than face up to a public relations disaster, NASA launches an unmanned rocket and blackmails the astronauts to fake a landing on Mars on a movie set. To keep them from spilling the beans, NASA schemes to bump them off, leading to the inevitable chase scene at the end. The real NASA must have thought it was good fun; the agency lent a hand in the film's making.

Bad Trips: 1980s to mid-'90s

After Viking, NASA was cooking, with three big projects underway: the Space Shuttle, the Hubble Space Telescope, and a billion dollar baby called Mars Observer. The latter was due to launch in 1988, but was delayed as ever more complicated instruments were added (including a high-resolution camera to photograph the Face). The Soviets, for their part, used the 1988 launch window to send **Phobos 1** and **Phobos 2** to study Mars' larger moon; contact was lost with Phobos 1 shortly after leaving Earth, while Phobos 2 managed to take a few photos and detect a faint outgassing, but before it could analyse the nature of the gas (most scientists think it was water) the Great Galactic Ghoul added it to its trophy collection.

In 1989, popular interest in space in general led President George Bush to call for a manned voyage to Mars. NASA totted up a wish list (including new launchers and spacecrafts, and the construction of a base on the moon as a first step) and handed in its estimate: $450 billion. The idea went no further; even Washington can occasionally get sticker shock.

Mars from Mars Observer

Then the troubles began. Launched with great expectations in April 1990, the **Hubble Space Telescope** was an embarrassing lemon, sending back fuzzy photos (until 1993, when astronauts on the Space Shuttle fixed it). Then one of the Space Shuttles blew up before the eyes of the world. In 1992, **Mars Observer** finally blasted off, taking with it high hopes of redeeming NASA in the public eye. In 1993, as it made its way to the Red Planet, a Mars Summit was held in Germany, with scientists from NASA, the European Space Agency, Russia, France and Japan, who all agreed to co-operate towards the goal of round-trip visits to Mars. But in August, three days before the Mars Observer was to arrive, it blew a fuel line and disappeared forever from NASA screens.

The tabloids had another field day; NASA's budget was slashed to ribbons. Mars was out of fashion, and all the brave ideas were quickly forgotten. Public support for the International Space Station to replace the sickly ageing **Mir** stagnated. Mars became something of joke, in bittersweet satires like Terry Bisson's *Voyage to the Red Planet* (1990), or the remake of *Invaders from Mars* (1986) or spoofs like *Lobster Man from Mars* (1989). Yet, two of Philip K. Dick's stories were adapted into blockbuster films: Ridley Scott's *Blade Runner* (1982), based on *Do Androids Dream of Electric Sleep?* (often rated the best sci-fi film ever) and *Total Recall* (1990), from *We Can Remember It for You Wholesale*, featuring over-the-top comic-book style special effects and Arnold Schwarzenegger, who saves the day by getting those ancient underground Martian atmosphere-making machines back on line.

Meanwhile the realistic possibilities of living on a terraformed Mars teased the imagination of writers even as NASA and the Russians floundered in disarray. Ian McDonald's *Desolation Road* (1988) offers a half real, half magical account of terraformed Mars; in 1992 Ben Bova creates a realistic scientific picture of a first manned mission in *Mars*, and Kim Stanley Robinson publishes his ultra-realistic *Red Mars*, the first of the superb trilogy about settling the planet. Then there's Greg Bear, who suggests that we can save ourselves much of the hassle by simply relocating Mars to a better neighbourhood (*Moving Mars*, 1993).

But Is there life, or at least SLIME on Mars?

Just by coincidence, 1976, the same year that the Vikings landed, was the year that the first parties of the Antarctic Search for Meteorites Program (ANSMET) went south to look for meteorites: for some reason, Antarctica seems to be the landing place of choice for space junk, and for us it's just as well. The search yielded examples that hit Earth 13,000 years ago, which matched other peculiar meteorites that fell to earth in the 19th century, known as shergottites, nakhlites and chassignites, or just SNCs. Similar SNCs were found on the Alan Hills in Antarctica. Thanks to the data supplied by Viking, it became evident in 1981 that the SNCs originated on Mars and had been hurled into space as *ejecta* by collisions with small asteroids.

In the 1990s, two discoveries reopened the book on the possibilities of life on Mars. In the basalt formations by the Columbia River in Washington state, Todd Stevens and Jim McKinley discovered what they called 'a subsurface lithoautotrophic microbial ecosystem', or SLIME for short—a highly ascetic ecosystem of bacteria living in hundreds of metres below ground, with neither sun nor thermal earth to warm them, and existing on nothing but a chemical action between the rock and water that generated natural gas-methane. The discoverers suggest that similar conditions could exist under the surface of Mars.

Then in August 1996, Dr David McKay and his co-workers at the Johnson Space Center in Houston stunned the world by announcing possible signs of Martian life. Fossils of

> Antarctica seems to be the landing place of choice for space junk

what appear to be microbacteria, a hundred times smaller than a hair's breadth, and chemical traces associated with life four billion years old were observed in Alan Hills 84001, an SNC the size of a potato. **AH 84001** was flung off the surface of Mars some 15 million years ago, and fizzled out in the snows of Antarctica 13,000 years ago. Fissures in the meteorite were found to contain carbonate globules, and inside these are tiny tubular structures that look like fossilized bacteria, as well as crystals of the minerals magnetite and iron sulphide, which are produced by some bacteria on earth, and polycyclic aromatic hydrocarbons (PAHs), organic molecules often formed when living organisms decompose.

A few months later in London, Colin Pillinger, Monica Grady and Ian Wright of the Open University and Natural History Museum—the scientists who had been the first to study the SNCs for fossils, and made the first discoveries that inspired David McKay to have a look—announced that they too had discovered similar smoking guns of life in two other meteorites. The strongest signatures in one of the British SNCs just happened to be very similar to those produced by methane-producing microbes on Earth.

With the news, every scientist on the planet wanted a piece of the rock, or at least a sliver, and it wasn't long before doubts poured in on all sides. Could the meteorites have become contaminated somehow during their long time on Earth? (PAHs have been found in the Antarctic ice, and in other meteorites found there.) And what about the temperature at which the possible cells in the rock were originally formed? Wouldn't it have been been way too hot for life? The McKay supporters so far have yet to be convinced and stick to their biological explanation; if researchers can find non-biological explanations for each of the three 'signs of life' the fact that all three are found together, they feel, still weighs in favour of life. Like so many Martian debates, this one remains inconclusive. But it spurred a new presidential interest in Mars, and cries of anguish from those directly involved that projects for Mars were severely underfunded.

The Pathfinder landing site

1996-97 Here we go again

With all the big excitement generated by AH84001 in August, everyone looked towards the next launch window, 1996, with high hopes. After yet another ambitious Russian launch, **Mars 96**, plunged into the Pacific, like a dagger into the heart of their space programme, a meaner, leaner NASA, under its new budget-minded director Daniel Goldin, sent up a pair of spacecraft: a lander called **Pathfinder**, and an orbiter, the **Mars Global Surveyor**. The string of bad Martian luck changed dramatically on the 4th of July when Pathfinder made a near perfect landing, right on target in the ancient flood plain near Ares Vallis, bouncing down on its industrial strength airbags. Pathfinder's rover, named **Sojourner** (after Sojourner Truth, a black woman who helped slaves escape from the south before the Civil War) was a cute remote-control buggy the size of a microwave oven on a skateboard, and it caught the public's fancy in a way that none of the big expensive probes ever managed; the photos sent back were immedately published on the NASA sites on the Internet, which registered over 100 million 'hits' the first day. Remote controlled from California like a toy (it

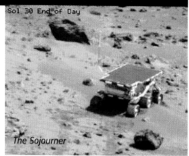

The Sojourner

had its practice runs on Earth in a 'sandbox'), it visited its neighbours in the 'Rock Garden': Barnacle Bill, Yogi, Mini Matterhorn, Mermaid Dunes and 'shook hands' with them, learning their composition. Originally designed to last a month, Pathfinder (now renamed the Carl Sagan Memorial Station) and Sojourner continue to recharge their batteries and function well, as we write this; Sojourner even performs wheelies to kick up the powdery dust and analyse the surface underneath.

Meanwhile the orbiter, the Mars Global Surveyor, arrived on 11 September for a two-year mapping and reconnaissance mission. One of its tasks will be to answer to the burgeoning 'Face on Mars' cottage industry, which is clamouring for the Cydonia features be re-photographed. NASA, apparently, intends to do just that. There's a scientific reason, in that the landforms of Cydonia are important in understanding Martian erosion, but NASA, dependent on taxpayer funding, can hardly duck public interest in the Face and the other features. NASA pleads that there may be technical problems: the cameras on the Surveyor may not be able to aim precisely at such a small area (3km square), or atmospheric conditions may be wrong, or a storm may occur when it passes over the site; the odds of snapping new pics of the Face are not overwhelming.

Long-term Stays: Colonies on Mars

Martian colonies have long been a favourite device of sci-fi writers, but their models are often spiked with intolerance, greed, lust for power, or madness; human nature, they warn you, just can't leave home without it. Ironically, however, once the Viking mission irrevocably proved that Mars had a freezing, poisonous, unpressurized, radioactive nature of its own, colonizing Mars has become a favourite fantasy of many scientists and engineers. The very difficulties of the task test their ingenuity, and the question of 'why bother?' a few years ago is more often now 'when?' Whatever else it accomplished, the Pathfinder Mission has renewed confidence in the old 'can-do' spirit, and now NASA is talking again about a manned mission, perhaps by 2011, for a fraction of the original estimate of $450 billion given to President Bush. And enthusiasm for colonies grows.

Probably the two most energetic boosters of colonies are Americans Dr Chris McKay of NASA, in charge of investigating past, present and potential life on Mars, and Dr Robert Zubrin, former senior engineer at Lockheed Martin and author of *The Case for Mars*, a practical step by step programme for going about it. Although still expensive (Zubrin estimates a first manned mission will cost $20 to $30 billion—a bargain, really, compared to the development of rum weapons like the B2), all scientists agree the effort would pay dividends in knowledge and the development of new technologies applicable at home as well as on another world.

Once it has been shown that we can get people to and from Mars, colonizing can begin. A handful of engineers and astronauts would arrive

Western humanist civilization as we know and value it today was born in expansion, grew in expansion, and can only exist in a dynamic expanding state.
Robert Zubrin, The Case for Mars

first to prepare the shelters—probably inflatable, air tight, pressurized structures, running on solar or nuclear power. Every two years more colonists will join them; eventually there will be greenhouses to grow food, underground housing (better protected from the radiation and cold) and stadium-sized domed structures, for recreation, gardens and raising animals. According to the *Mars Habitation 2057* plan put forth by Japanese scientists Dr Yoji Ishikawa, Takaya Ohkita and Yoji Amemiya, the colony could expect to have a population of 50,000 by the year 2100. Venturing outside would always mean donning pressurized space suits. This is a favourite plan of Mars Greens (or Reds, in this case) who believe we should leave Mars' environment as untouched as possible.

Other plans are more ambitious, reaching a climax with *Worldhouse*, created by Richard L.S. Taylor of the British Interplanetary Society; this calls for covering 80 per cent of the surface of Mars over the next 250 years

Julian Baum's artist's impression of Worldhouse

with a two-mile high, transparent gas-tight roof, supported by pylons and skyscrapers, each capable of housing 500,000 souls. Oncoming comets or meteorites would be nuked or diverted like the Scuds in the Gulf War with land and air based 'Star Wars' techniques.

But what most fascinates and repels Marsologists is the idea of terraforming, making the Red Planet more like our blue one—with warm, breathable air, liquid water, and a proper atmosphere—restoring the Mars of some four billion years ago. Terraforming is not something that immediately appeals to anyone with a short attention span—depending on who you talk to, the full process is estimated to take between 1000 and 100,000 years.

The first phase of terraforming would raise the average temperature of Mars to about 32°F (0°C), allowing water to exist as liquid. Chris McKay wants to do this by the very same technique that is currently heating Earth's climate: building nuclear and solar-powered factories spitting out massive quantities of greenhouse gases—chlorofluorocarbons (CFCs) and perfluorides—which are composed of the elements that conveniently exist in the salts of Mars' soil: carbon, fluorine and chlorine.

Other suggestions to speed up the process include positioning giant mirrors in orbit around Mars to direct sunlight towards the poles, and spreading soil over the polar caps to reduce the loss of solar energy reflected back into space. Raising the temperature, even slightly, is the key to terraforming. Melting the poles would release CO_2 gases (now in the form of dry ice), and more could be released from rocks in the soil, beefing up the atmosphere to the point where hardy plants could grow, and we could get around without pressure suits. It would also release the water locked in the permafrost, and bring lakes and seas back to the Martian landscape.

Using models based on Earth's climate, Zubrin and McKay believe it possible to create a runaway greenhouse effect, that would, after one or two hundred years of pumping out 900 tonnes of gasses per hour, raise the average temperature of Mars to -5°C, a warmer moister Mars with flowing water in places. The terraformers' main worry is that Mars may have insufficient quantities of CO_2, water and nitrogen (the most doubtful—much of it might have blasted off the atmosphere in solar winds). One of the tasks of the next decade of NASA missions is to determine just how must of these ingredients are there. The loss of the Russian Mars 96 was especially galling to the terraformers, as it was meant to dig for nitrates in the soil, where much of Earth's own nitrogen is stored.

Recalling that Mars was originally warmed billions of years ago by the energy released by the impact of millions of meterorites, McKay and Zubrin also suggest lassooing ammonia-rich asteroids from the outer solar system and guiding them to crash into Mars; the ammonia that would be released in the atmosphere would also act as a greenhouse gas. Britain's Dr Martyn Fogg suggests, more controversially, that we set off tremendous nuclear explosions under the surface of Mars to release the underground water to warm the planet. Kim Stanley Robinson, in his colonizing trilogy *Red Mars, Green Mars, Blue Mars* has the colonists hijack giant ice water asteroids that melt as they crash into the atmosphere and up the pressure on Mars with their water vapor. Genetically engineered micro-organisms could also produce greenhouse gases (like methyl chloride) in a special recipe that would also eventually provide a shield against radiation.

Once Mars was sufficiently warm, the much longer terraforming second phase could begin. Micro-organisms—cousins of SLIME—living on water and trace elements would use sunlight to reduce the carbon dioxide in the atmosphere, and boost the oxygen to a level we could breathe. This process happened on Earth a few billion years ago; on Mars it may take 100,000 years.

So start now, urge McKay and Zubrin. New international cooperative efforts between scientists and government agencies have already begun. Meanwhile, Zubrin and others (notably the Space Frontier Foundation) believe that getting private enterprise involved in the space race may speed up efforts considerably: one idea is for Washington to offer 'Mars Prizes' for various technological breakthroughs, leading up to a $20 billion prize for whoever first sent a crew to Mars and back. In *The Case for Mars*, Zubrin often compares the new frontier of America in the last century to the Martian frontier at the beginning of the new millennium. His ultimate, and perhaps most winning argument for such a new interplanetary frontier is humanistic: 'America created a new standard for the treatment of humans. The same could be true with Mars.'

Global Surveyor launch

Marsology
Further Reading

If this little guide to Mars has made you keen to find out more before you go, it has done its duty. For more Martian bibliographies (fiction, sci-fi, research papers, short stories, going back to the 19th century), check out the websites compiled by Gene Alloway, Senior Associate Librarian at the University of Michigan: http://www-personal.engin. umich. edu/~cerebus/mars/

Non-Fiction

- **Batson, R. M., Bridges, P. M. and Inge, J. L.,** *Atlas of Mars*, NASA Scientific and Technical Information Branch, 1979. The first complete Martian atlas, with Viking and Mariner pictures.
- **Bear, Greg,** *Moving Mars*, Tor, 1993. You gotta move it, move it—one man's solution to the planet's hostile climate.
- **Caidin, Martin, Barbee, Jay and Wright, Susan** *Destination Mars: In Art, Myth and Science*, Penguin Studio, 1997. Perceptions of Mars through history, with pretty pictures.
- **Carlotto, Marc J.,** *The Martian Enigmas: A Closer Look: The Face, Pyramids, and Other Unusual Objects on Mars*, North Atlantic Books, 1997. Self explanatory.
- **Carr, Michael H.,** *The Surface of Mars*, Yale University Press, 1981. A dry and technical tour of the landscape, but tons of sound fact and wonderful photos of the surface.
- **Catermole, Peter,** *Mars, the Story of the Red Planet*, Chapman and Hall, 1993. Everything about the planet's development and geology, but not very accessible to non-geologists.
- **Clarke, Arthur C.,** *The Snows of Olympus: A Garden on Mars* Gollancz, 1994, W.W. Norton, 1995. Grand old man of sci-fi discusses the prospects for terraforming Mars.

- **Collins, Michael,** *Mission to Mars*, Grove 1990. The only Mars book by a real astronaut, on future colonies; suggests sending married couples only for stability.
- **Goldsmith, Donald,** *The Hunt for Life on Mars*, E. P. Dutton, 1997. The first book on the AH84001 meteorite, but not the last. Explains what the fuss was all about.
- **Hoagland, Richard C.,** *The Monuments of Mars: A City on the Edge of Forever*. North Atlantic Books, 1987. A thoughtful case for intelligent life in Cydonia—not necessarily a convincing one.
- **Hoyt, William Graves,** *Lowell and Mars*, University of Arizona Press, 1976. A full account of the great man and his grand delusion.
- **Lovelock, James and Allaby, Michael,** *The Greening of Mars*, Andre Deutsch, St. Martin's/Marek, 1984. The co-creator of the Gaia hypothesis and a science-fiction writer combine for a utopian view of Mars' future: peace, veggies, airships, and a Martian population that rapidly evolves into a separate species.
- **McDaniel, Stanley V.,** *The McDaniel Report-On the Failure of Executive, Congressional and Scientific Responsibility in Investigating Possible Evidence of Artificial Struction*, North Atlantic Books, 1994. Cydonia Face and conspiracy theories; newsletter on the web.
- **Sheehan, William,** *The Planet Mars*, University of Arizona Press, 1996. One of the best accounts of early Mars observation.
- **Smith, A E,** *Mars, The Next Step*, Adam HIlger, 1989. Mars missions and colonization prospects.
- **Stoker, Carol R., and Emmart, Carter, eds.,** *Strategies for Mars*, Univelt 1996. Essays and cost/benefit analysis

- **Viking Lander Imaging Team,** *The Martian Landscape*, NASA Scientific and Technical Information Office, 1978. Original report on the surface from the Viking mission, lots of pictures.
- **Washburn, Mark,** *Mars at Last*, Putnams, 1977. Written right after Viking, good on the Viking and Mariner missions, pre-scientific beliefs and Martian sci-fi.
- **Wilford, John Noble,** *Mars Beckons*, Knopf, 1990. Of the spate of recent books that cover the whole Martian schtick from the Greeks to prospects for colonization, this is one of the best-written and most informative.
- **Zubrin, Robert** (with Richard Wagner), *The Case for Mars*, Free Press, 1996. This book might make history, advocating that we start right now, and explaining how to do it. Written with depth and passionate intensity.

Fiction

- **Anderson, Kevin J.,** *War of the Worlds: Global Dispatches,* Bantam/Spectra, 1996. Anthology of short stories, of Wells's War of the Worlds seen through the eyes of late 19th-century celebrities. Good fun.
- **Baxter, Stephen,** *Voyage*, Harper Collins, 1997. Realistic if tragic novel that presupposses that JFK was never assassinated, and oversaw the first manned mission on Mars in 1986.
- **Bisson, Terry,** *Voyage to the Red Planet*, William Morrow & Co, 1993. Inspired by Burroughs, satire and fantasy—an attempt by has-beens to make money off a trip to Mars
- **Bova, Ben,** *Mars*, Bantam, 1992. A realistic account of the first manned mission to Mars
- **Bradbury, Ray,** *The Martian Chronicles*, also published as *The Silver Locusts,* many reprintings, from 1951; reprinted by Vintage, 1995 and others; a classic, poetic and evocative. We are the alien intruders who selfishly destroy a better world, and are destroyed in turn.
- **Burroughs, Edgar Rice,** between 1917–1924, wrote the Barsoom series of space operettas: *A Princess of Mars, John Carter of Mars, Swords of Mars, Synthetic Men of Mars, The Chessmen of Mars, Master Mind of Mars*. Last reprints of the lot were by Ballantine in the early 1960s; New English Library reprinted the best in 1972. They're awful--but they inspired most of the better sci-fi writers who came after.
- **Butler, Jack,** *Nightshade*, Atlantic Monthly Press, 1989. Human colonies on Mars, and vampires, too.
- **Compton, David,** *Farewell Earth's Bliss*, Borgo Press, 1979. Saga of Mars used as a dump for maladjusted earthlings.
- **Deitz, William C.,** *Mars Prime*, ROC, 1992 . Mars as a prize for greedy bounty hunters.
- **Dick, Phillip K.,** *Martian Time Slip*, first printed by Ballantine, 1964, reprinted by Vintage, 1995. T*he Three Stigama of Palmer Edlritch*, 1965; *Do Androids Dream of Electric Sheep?*, Del Rey, 1996 (among many editions); *We Can Remember it For You Wholesale* ; Collection of Stories, Volume 2, Citadel, 1990.
- **Dozois, Gardner, ed**. Isaac Asimov's *Mars*, Ace, 1991. Short stories by various sci-fi writers.
- **Farrer, Mick,** *The Red Planet*, Del Rey, 1990. Military-enhanced humans on Mars.
- **Heinlein, Robert,** *Red Planet*, Stranger in a Strange Land, 1969, reprinted by Ace, 1995, among many editions. Human raised on Mars returns to earth. First genuine sci-fi novel to make the *New York Times'* bestseller list
- **McCauley, Paul J.,** *Red Dust*, Gollancz, 1993, William Morrow 1994. Seven hundred years from now, a lonely exile on a forced labour farm on a politically repressed Mars rescues a space alien; meanwhile ruthless tyrants are plotting to turn a terraformed Mars into dust.
- **McDonald, Ian,** *Desolation Road*, Bantam, 1988. Author's first novel on the terraforming of Mars, a mixture of fantasy and reality.
- **Pike, Christopher,** *The Season of Passage,* 1992, Hodder and Stoughton, London. Horror on Mars.

- **Pohl, Frederik**, *Man Plus*, Gollancz, 1976; Random House, 1976. Nebula award winner, on turning into a cyborg to cope with the harshness of Mars; the sequel, Mars Plus, Baen Books,1994, written with Thomas T. Thomas, leaps ahead 50 years, when the cyborgs and misfits have to face a cranky computer; also *The Day the Martians Came* (1988), a romance between aliens and earthlings
- **Robinson, Kim Stanley**, *Red Mars* (1992), *Green Mars* (1993), *Blue Mars* (1996) HarperCollins/Voyager; Bantam Spectra. The most scientifically realistic trilogy on the terraforming of Mars, and the opposition to the project on Earth. Also see Robinson's *Icehenge Futura*, 1984, and *The Memory of Whiteness*, TOR, 1985.
- **Shatner, William**, *Man O'War*, Putnam, 1996. Diplomat posted to Mars—turned into a slave planet to provide for billions on Earth—and consequent revolt.
- **Silverberg, Robert**, *Lost Race of Mars*, Scholastic Book Services, 1960; *Mars of Time* (also published as *Vornan-19*) Sidgwick & Jackson, 1968. Solid Mars fiction for the kids; *Mars of Time* was a Nebula Nominee.
- **Stabenow, Dana**, *Red Planet Run*, Ace, 1995. Probably the only Mars book by an Alaskan, a story with space pirates and the obelisk responsible for all the lost missions
- **Turner, Frederick**, *A Double Shadow*, Putnam 1978. One of America's best known poets dips into sci- fi, comparing life on Mars, the frontier, and the decaying Old Earth World
- **Wellman, Mac, Annie Salem**, *An American Tale*, Sun and Moon, 1996. Eccentric but pleasurable. Smalltown Ohio boy in love with Annie Salem ends up having a showdown with Ku Klux Klan on Mars, and much much more.
- **Wells, H. G.**, *The War of the Worlds*, New American Library, 1996, among many, many editions. The first and best account of Martian invaders.

Further Listening

- *Attack of the Radioactive Hamsters From a Planet Near Mars*, by Weird Al Yankovic
- *BB on Mars*, by Alice Cooper on the album 'Pretties for You'.
- *Bird Dream of Olympus Mons*, by the Pixies, on 'Trompe Le Monde'. Mars' gigantic volcano.
- *Deimos and Phobos*, by Steve Lyon. A Martian misses his double mooned home.
- *From the Mars Hotel*, (1974) the Grateful Dead. Spaced out.
- *Guitars on Mars* (1997) Ambient Compilation
- *Here Come the Martian Martians* (1976) by Jonathan Richman & the Modern Lovers. They come, but can't cope in our capitalist society.
- *Invaders from Mars* by Crack the Sky, from the album 'Animal Notes'.
- *Life on Mars?* (1972) by David Bowie The most famous Martian anthem.
- *I want my Baby on Mars*, by Bow Wow Wow, Malcolm McLaren's proteges.
- *The John Carter Songbook*, 1979–85) Sten Hanson took clues about Martian music from Burrough's and added vocals with a computerized synthesizer.
- *Little Green Men*, by Steve Vai, from his album 'Flex-Able'.
- *Live... from the Tea-Rooms of Mars*, by Landscape on the album 'From The Tea-Rooms of Mars...to The Hell Holes of Uranus'.
- *Man from Mars*, by Blondie, from 'Rapture'
- *Mars* Audiac Quintet (1994) album by Stereolab.
- *Mars Needs Guitars*, by the Hoodoo Gurus
- *Mars Needs Women* by Fleck, Bela and the Flecktones
- *Martian Cowboy*, by Toyah, on the album 'Love is the Law'.
- *The Martian Hop*, by the Ran Dells (1963). The Martians get down. A classic.

- *May The Cube Be With You*, by Thomas Dolby. A Martian song on the album 'Aliens Ate My Buick'.
- *The Planets*, Gustav Holst's symphonic journey (1920) through the solar system; the section on the Red Planet, called 'Mars, the Bringer of War' is typically loud and noisy.
- *Stuart*, by the Dead Milkmen: from 'Beezelbubba': song about a nut in a trailer park who thinks that the gay contingent of Des Moines, Iowa, is secretly building landing strips for homosexual Martians.
- *Uncle Sam's on Mars* by Hawkwind, the all-time champ sci-fi fantasy band.
- *Venus and Mars Reprise* by Wings, from 'Venus and Mars'.
- *War of the Worlds* Radio Play (1994) Orson Wells. The original, uncut 1938 radio broadcast on CD.
- *War of the Worlds*, (1978) by Jeff Wayne. H.G. Wells set to music, with Richard Burton doing narration. Hear it at http://www.war-of-the-worlds.com/.

Further Watching

- *A Message from Mars* (1912) US silent film. Man heading up wrong path of life receives sound advice from disinterested Martian.
- *Aelita: Queen of Mars* (1924) In this silent Soviet film directed by Jakov Protazonov, the engineer-hero builds a rocket, kills his do-gooder wife, and blasts off to Mars, which turns out to be an opressive capitalistic monarchy ripe for revolution. The engineer falls in love with Aelita, the Queen of Mars, then wakes up—like Dorothy in Oz. It was all a dream.
- *Flash Gordon: Mars Attacks the World* (1938) starring Buster Crabbe, Jean Rogers, Frank Shannon and Charles Middleton. Mars is sucking the nitrogen out of Earth's atmosphere, and Flash Gordon, girlfriend Dale Arden, and genius Dr. Zarkov zoom off the red planet to round up the usual suspects. Here they find Ming the Merciless stirring up strife between the Clay People of Mars and Prince Barin of the planet Mongo, hoping to start a battle that will enable him to conquer the universe when no one's looking. But the Clay People make peace with Barin and Flash Gordon, and Ming is defeated again (He escapes, and for that matter he's still at large).
- *Rocketship X-M* (1950) starring Lloyd Bridges, Hugh O'Brian and Morris Ankrum. Five moon-bound astronauts have technical troubles and get detoured to Mars, only to find an advanced civilization in ruins after being nuked. The only Martians who survived the holocaust have regressed into primitive savages. Three astronauts live to escape, but unfortunately their spacecraft crashes before earth receives the anti-atomic war message.
- *Flight to Mars* (1951) starring Carmon Mitchell, Arthur Franz, and Morris Ankrum. The first Martians in colour. Scientists and a reporter go to Mars and learn that Martians look just like us. They're also pretty sneaky, and want to steal the humans' rocket to invade Earth for its 'corium'. The day is saved thanks to pro-Earth Martian underground.
- *Red Planet Mars* (1952) Starring Peter Graves, Andrea King, Morris Ankrum. A hysterically stupid movie from the McCarthy era where earthlings receive TV messages from Mars and learn that the Red Planet is a paradise in direct contact with God. When the godless Russians realize this, their government collapses and the world is a better place.
- *Invaders from Mars* (1953) One of the classics, done from a child's point of view. The young hero sees strange lights landing behind a hill and soon after notices that his parents, and all the adults around him begin to behave in a very odd manner. This is because they have Martian gadgets stuck in their necks.

- *Abbott and Costello Go to Mars* (1953) The rocket, which Abbott and Costello accidentally ignite, goes to New Orleans just in time for *Mardi Gras*. There bank robbers force them to fly to Venus, where the women (played by beauty pageant contestants) have expelled all the men.
- *The War of the Worlds* (1953) Directed by Byron Haskin, starring Gene Barry; winner of an Academy Award for special effects. Red Planet baddies bite the dust, this time in California of course, in a film adaptation made at the height of the Cold War.
- *Devil Girl from Mars* (1954) starring Patricia Laffa, Hazel Court, Hugh McDermott and Adrienne Corri. Mars needs men. Mischevious alien in leather and her killer robot sidekick descend on earth to bring them back.
- *The Angry Red Planet* (1959) starring Gerald Mohr, Nora Hayden, Les Tremayne. Astronauts on Mars vs. unwelcoming locals, gelatinous crud, and other nuisances.
- *Santa Claus Conquers The Martians* (1964) starring John Call, Pia Zadora, Jamie Farr. Here the Martians play Scrooge: they hate to see earth kiddies being happy so they swipe Santa and take him to Mars. But the next thing you know sugar plums are dancing in their alien heads.
- *Robinson Crusoe on Mars* (1964) dirercted by Byron Haskin. After crashing on Mars, an astronaut tries to survive on Mars and keep his sanity with only a monkey for company, until he meets a Martian Friday
- *Planet of Blood* (1966) starring Basil Rathbone, John Saxon, Dennis Hopper, and Florence Marly. Astronauts visiting Mars find the debris of a spaceship from another planet and make the mistake of taking home the sole survivor: a green bloodsucking female alien.
- *Mars Needs Women* (1966) starring Yvonne Craig. Lonely Martian males embark on an interplanetary possee to round up some ladies. Critics warn that you might think this is one of the films that's so bad it's good, but it isn't.
- *Capricorn One* (1978) Wicked NASA fakes its first manned mission to Mars to prevent a public relations disaster, then tries to knock off the astronauts.
- *Invaders from Mars* (1986) starring Karen Black, Hunter Carson, Timothy Bottoms. Remake of the 1953 film.
- *Lobster Man from Mars* (1989) In a plot reminiscent of Mel Brooks' *The Producers*, a film producer who needs a dud flick for a tax write-off buys a film student's goofy project about Martians who send the nefarious Lobster Man over to steal Earth's air, only to be thwarted by a mad scientist, a girl, and an army colonel. The film is an enormous success and the producer goes to jail.
- *Total Recall* (1990) starring Arnold Schwarzenegger, Sharon Stone, directed by Paul Verhoeven. Disturbed by a recurrent dream about a trip to Mars, Schwarzie buys an implanted memory at Rekall Inc. for a vacation, but it shorts out and he remembers that he was a secret agent battling villains on Mars. The fight recommences, and after massive gratuitous violence, Mars' underground nukes blow and its water is liberated.
- *Mars Attacks!* (1996) Directed by Tim Burton, starring Glenn Close, Jack Nicholson, Natalie Portman, Pierce Brosnan, Annette Bening, Danny Devito, Rod Steiger, Tom Jones. Based on the goofy Topps bubble gum card series, really horrid Martians attack Earth and wreck havoc before we send them packing.

...rEady for lift-off!